## Advances in Optics and Optoelectronics

**Series Editor**

Perry Ping Shum, Southern University of Science and Technology, Fanling, Hong Kong

The Advances in Optics and Optoelectronics series focuses on the exciting new developments in the fast emerging fields of optics and optoelectronics. The volumes cover key theories, basic implementation methods, and practical applications in but not limited to the following subject areas:

AI Photonics
Laser Science and Technology
Quantum Optics and Information
Optoelectronic Devices and Applications
Fiber-Based Technologies and Applications
Near-infrared, Mid-infrared and Far-infrared Technologies and Applications
Biophotonics and Medical Optics
Optical Materials, Characterization Methods and Techniques
Spectroscopy Science and Applications
Microscopy and Adaptive Optics
Microwave Photonics and Wireless Convergence
Optical Communications and Networks

Within the scopes of the series are monographs, edited volumes and conference proceedings. We expect that scientists, engineers, and graduate students will find the books in this series useful in their research and development, teaching and studies.

Vanita Bhardwaj · Sonal Khurana · Rekha Bhardwaj
Editors

# Photonics and Optoelectronics in Industry 5.0

Springer

*Editors*
Vanita Bhardwaj
School of Engineering and Technology
Vivekananda Institute of Professional Studies-Technical Campus
New Delhi, Delhi, India

Sonal Khurana
School of Engineering and Technology
Vivekananda Institute of Professional Studies-Technical Campus
New Delhi, Delhi, India

Rekha Bhardwaj
Department of Applied Sciences
Bharati Vidyapeeth's College of Engineering
New Delhi, Delhi, India

ISSN 2731-6009 ISSN 2731-6017 (electronic)
Advances in Optics and Optoelectronics
ISBN 978-981-96-7593-7 ISBN 978-981-96-7594-4 (eBook)
https://doi.org/10.1007/978-981-96-7594-4



This Springer imprint is published by the registered company Springer Nature Singapore Pte Ltd.
The registered company address is: 152 Beach Road, #21-01/04 Gateway East, Singapore 189721, Singapore

*With deep gratitude, we dedicate this book to our families, whose unwavering support, patience, and encouragement have been our greatest strength throughout this journey. Their love and belief in us have inspired us to persevere and bring this book to life.*

*We also extend our heartfelt appreciation to our fellow authors, whose valuable contributions, expertise, and dedication have enriched this work. Their collaborative spirit and commitment have been instrumental in making this book a reality.*

*A special note of thanks to our friends and colleagues, whose insights, encouragement, and support have motivated us throughout this endeavour.*

*Above all, we are deeply thankful to everyone who stood by us, inspiring us to pursue this project with passion and dedication.*

*Dr. Vanita Bhardwaj*
*Dr. Sonal Khurana*
*Dr. Rekha Bhardwaj*

# Preface

The transition from industry 4.0 to 5.0 signifies an evolution toward more human-centric, sustainable, and intelligence of industry landscape. Integration of photonics and optoelectronics has emerged as significant facilitator of this evolution as the emphasis shifts from mere automation and efficiency to creativity, collaboration, and sustainability.

Photonics and optoelectronics, with their unparalleled strengths in energy efficiency, precise sensing, proficient manufacturing, are rapidly establishing themselves as essential tools in achieving industry 5.0 goals. The main objective of this book, *Photonics and Optoelectronics in Industry 5.0*, is to provide an in-depth analysis of these advanced technologies and their applications in the dynamic environment framework.

The content of book is organised into several parts and begins by providing strong foundation of fundamental concepts, followed by in-depth real-world applications of them and challenges. The transition from theoretical knowledge to real-world applications highlights how photonics and optoelectronics have been integrated into supply chain management, sophisticated manufacturing process, and renewable energy system.

The chapters include:

Chapter 2 "**Introduction to Photonics and Optoelectronics in Industry 5.0**"—An overview of the transition from Industry 4.0 to Industry 5.0 and the role of photonics and optoelectronics in enabling this shift. Chapter 3 "**Fundamentals of Photonics and Optoelectronics in Modern Industrial Processes**"—A detailed exploration of the essential concepts and technologies that form the backbone of photonics and optoelectronics. Chapter 4 "**Integration of Photonics and Optoelectronics in Industry 5.0**"—Examination of how these technologies are embedded into various aspects of Industry 5.0, enhancing communication, manufacturing, and human–machine interaction. Chapter 5 "**Implementing Photonics in Advanced Manufacturing Techniques for Industry 5.0**"—Insights into laser machining, precision manufacturing, and other advanced techniques enhanced by photonic technologies. Chapter 6 "**Real-Time Monitoring and Quality Control Using Photonic Technologies in Industrial Environments**"—Exploration of how optical sensors and imaging systems contribute to improved consistency, reliability, and safety. Chapter 7 "**Role of Photonics in Enhancing Additive Manufacturing Techniques and Process**"—Analysis of how photonics advances additive manufacturing with improved accuracy and efficiency. Chapter 8 "**Green Photonics Enhancing Supply Chain Management**"—Discussion on how photonics aids in achieving eco-friendly practices through energy optimization and carbon footprint reduction. Chapter 9 "**Innovative Photonics Technologies for Renewable Energy Systems in Industry 5.0**"—Examination of the role of photonics in enhancing renewable energy systems, including solar harvesting and smart grids. Chapter 10 "**Challenges and Opportunities in Photonics Integration for Industry 5.0**"—Identification of the obstacles to photonic integration and the opportunities for innovation and growth. Chapter 11 "**Emerging Trends and Future Directions in Photonics and Optoelectronics for Industry 5.0**"—A forward-looking perspective on the potential advancements and innovations in the field.

The collective efforts of authors, editors, and researchers have contributed to making this book a valuable resource for understanding the evolving role of photonics and optoelectronics in Industry 5.0. It is intended to serve as both a comprehensive reference for established researchers and a guide for those new to the field, fostering further innovation and exploration.

We extend our sincere gratitude to all the contributing authors whose expertise and dedication have been instrumental in the completion of this work. Their contributions have significantly enriched the quality and depth of this volume.

This book represents a collaborative endeavour aimed at bridging the knowledge gap between theory and application. The journey of integrating photonics and optoelectronics into Industry 5.0 is ongoing, and we hope this work inspires continued research and practical advancements in the field.

New Delhi, India

Vanita Bhardwaj
Sonal Khurana
Rekha Bhardwaj

# Acknowledgments

We would like to express our sincere thanks to everyone who helped make this book, *Photonics and Optoelectronics in Industry 5.0*, possible. Without their support, this work would not have been completed.

First, we want to thank the contributors and authors for sharing their expertise and knowledge. Their input has been essential in making this book informative and up-to-date on the Integration of *Photonics and Optoelectronics in Industry 5.0.*

We also appreciate the guidance and encouragement from our colleagues, mentors, and advisors. Their advice helped shape the content and quality of the book.

A big thank you goes to the team at *Springer Nature*, who provided excellent support throughout the publishing process. Their work in editing, formatting, and production has been invaluable.

Lastly, we would like to thank our families and friends for their ongoing support and patience while we worked on this book. Their encouragement was a great source of strength.

We hope this book will be helpful to everyone interested in the combination of optical fibre sensors and AI, and that it will inspire more research and developments in this exciting area.

# Acknowledgements

# About This Book

*Photonics and optoelectronics in Industry 5.0* is a detailed guide that explains how photonics and optoelectronics are helping industries move from Industry 4.0 to Industry 5.0. Industry 5.0 focuses on making technology more human-centred, intelligent, and eco-friendly. This book shows how photonic technologies are being used to improve accuracy, efficiency, sustainability, and collaboration between humans and machines. This book is a valuable resource for researchers, professionals, and students who want to understand how photonics and optoelectronics are shaping the future of industries. It aims to provide practical knowledge and insights to help readers apply these technologies effectively in Industry 5.0.

As part of the series *Advances in Optics and Optoelectronics*, this book aims to bridge the gap between research and real-world applications, encouraging innovation and sustainable growth.

We thank all the authors who have contributed their knowledge and expertise to make this book a useful and informative resource. Their efforts have made this work possible, and we are grateful for their valuable contributions.

# Contents

# Chapter 1
# Evolution of Industry 5.0 from Industry 4.0

**Mohd Javaid and Abid Haleem**

**Abstract** In the context of Industry 5.0, the fusion of human expertise with cutting-edge technologies such as artificial intelligence (AI), augmented reality (AR), virtual reality (VR), robots, big data analytics, the Internet of behaviours (IoB), and other innovative technologies is of the utmost importance. Photonics and optoelectronics are pivotal in facilitating high-speed data transmission and the implementation of AR and VR systems, which will enhance human–machine interactions. Additionally, devices like solar cells that use optoelectronics contribute to reducing energy consumption and carbon footprint in industrial processes. Industry 5.0 heavily depends on the technical advancements offered by photonics and optoelectronics, as they promote energy efficiency and the integration of smart devices, thereby contributing to the establishment of responsive and adaptable production environments. This chapter briefly discusses Industry 5.0 and the evolution of Industry 4.0 to Industry 5.0. Principal components, technologies, and application areas of Industry 5.0 are discussed. Industry 5.0 presents an encouraging production era where the synergy between humans and modern technology drives efficiency, sustainability, and creativity, with photonics and optoelectronics playing indispensable roles in this transformative journey.

**Keywords** Industry 5.0 · Industry 4.0 · Evolution · Technologies

## 1.1 Introduction

Industry 5.0 technologies are robust, sustainable, and focused on the needs of people. People will have limitless options for growth and development in the workplace of the future, and industrial resilience will lay the groundwork for flexibility to constantly shifting market conditions. Industry 5.0 holds great significance as it presents a chance to establish a manufacturing sector that is both compassionate and sustainable [1–3]. Industry 5.0 contributes to creating more satisfying and rewarding

M. Javaid (✉) · A. Haleem
Department of Mechanical Engineering, Jamia Millia Islamia, New Delhi, India
e-mail: mjavaid@jmi.ac.in

 
V. Bhardwaj et al. (eds.), *Photonics and Optoelectronics in Industry 5.0*, Advances in Optics and Optoelectronics, https://doi.org/10.1007/978-981-96-7594-4_1

employment for people and better working conditions by emphasising the human element more in the production process. It also presents the possibility of developing adaptable and efficient production procedures to better adjust to shifting consumer demands and supply chain interruptions. Industry 5.0, instead of being viewed as a logical development from Industry 4.0, is based mainly on the concept of humans and machines coexisting together [4, 5].

Industry 5.0 tackles the problem of doing away with human involvement while employing digital technologies. It also combines the idea of fusing human creativity with efficiency through technology. It plays a critical role in streamlining fabrication processes, which improves creativity, lowers waste, and saves money. AI dramatically improves quality control through machine learning and picture recognition. It can reduce expensive non-operational risks by forecasting machinery or equipment breakdowns by analysing past data and present circumstances [6, 7]. AI-powered monitoring systems are also excellent at quickly identifying risk indicators and implementing precautions to avoid mishaps. This represents a change from the past focus on economic value to a holistic viewpoint prioritising societal worth and well-being. Although comparable concepts have been studied in the past, such as corporate social responsibility, Industry 5.0 brings a fresh perspective on prioritising people and the environment before profits [8–10].

Industry 5.0 can now focus on what technology can do for workers rather than what can be done by them with new technologies. Realising that while robots are excellent at accuracy, they lack the human ability to think critically and creatively. The personalisation of human/machine interactions is made possible by embedded sensors, actuators, and machine learning technologies [11, 12]. This allows cobots, to adapt seamlessly. Industry 5.0 holds great significance as it presents a chance to establish a manufacturing sector that is both compassionate and sustainable. It may create more satisfying and rewarding employment and better working conditions by emphasising the human element more in production. It also presents the possibility of developing more adaptable and efficient production procedures to better adjust to shifting consumer demands and supply chain interruptions [13, 14]. This chapter's research objectives are as follows:

***Objective 1***: to provide an overview of Industry 5.0 and examine how it evolved from Industry 4.0 to Industry 5.0;
***Objective 2***: to outline the key distinctions between Industry 4.0 and Industry 5.0;
***Objective 3***: to examine the important elements of Industry 5.0 and provide an overview of its key technologies;
***Objective 4***: to list and briefly describe the Industry 5.0 application areas.

## 1.2 Industry 5.0

Industry 5.0 aims to increase production efficiency by collaborating between humans and robots. The main emphasis of Industry 5.0 is cobots, which can transform how people interact with technology in industrial settings. These cobots enable smooth

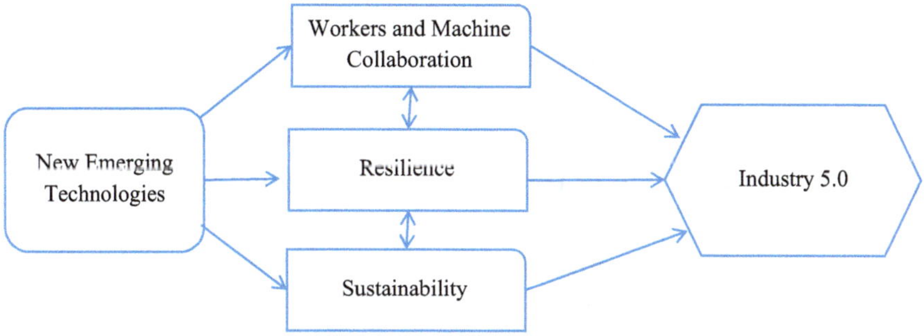


**Fig. 1.1** Process steps of Industry 5.0

collaboration and offer several benefits, such as protection, adaptability, and, most significantly, agility [15–17]. They are designed to recognise various indications from their environment thanks to multiple sensors, including force/torque, vision, and infrared (IR) sensors. They can detect body heat, detect variations in force and torque, and visually distinguish items and people in their workspace. Both can operate together in a safe and peaceful environment because of their extensive sensing capabilities.

The process steps of Industry 5.0 are shown in Fig. 1.1.

Industry 5.0 is predicted to usher in a new age of human–robot co-working dynamics, marked by maximum automation and little human involvement. The transformation of current smart factories into lights-out factories will be made possible in large part by this partnership. These will be made to support a fully robotic workforce with very little help from employees [18–21].

## 1.3 Evolution of Industry 4.0 to Industry 5.0

Several revolutionary revolutions have revolutionised the industrial landscape, each bringing about a significant advance in producing and delivering goods and services. First came Industry 1.0, which used steam and water power to mechanise processes. Next came Industry 2.0, which used electricity to power assembly lines and mass manufacturing. Smart factories were made possible by integrating cyber-physical systems, the IoT, and AI in Industry 4.0, which emerged from computer advances and automation in Industry 3.0. We are beginning the era of Industry 5.0, that prioritises sustainability, more customised manufacturing methods, and human–machine collaboration. The smooth integration of cutting-edge technologies characterises Industry 5.0. It heralds a new age of convergence between the digital, biological, and physical domains, paving the way for ground-breaking discoveries that might completely transform markets, economies, and society [22–26]. Industry 5.0 aims to create a more harmonious interaction between humans and technology, whereas Industry 4.0 brought about the digitisation and interconnectedness of many systems.

In addition to addressing urgent global issues like inequality and climate change, this new era aims to advance sustainability [27, 28].

## 1.4 Main Distinction of Industry 4.0 and Industry 5.0

The manufacturing and industrial environment dramatically changed with the migration from Industry 4.0 to Industry 5.0. Both revolutions have modern technology at their core but differ regarding priorities, methods, or end objectives. Industry 5.0 emphasises human–machine cooperation more than ever, with technology that enhances rather than replaces human talents [29, 30]. Utilising cutting-edge technology like robots, AI, and the IoT, it builds manufacturing systems that are more adaptive, flexible, and sustainable. It strongly emphasises environmental stewardship, using production techniques that reduce waste and pollution while maximising the use of natural resources. Employee education and training should be prioritised to equip staff members with the new skills and knowledge needed for Industry 5.0 and prepare them for the latest procedures and technologies [31–33]. Table 1.1 briefs the main distinction of Industry 4.0 and Industry 5.0.

Industry 5.0 embodies a human-centric strategy prioritising cooperation, customisation, sustainability, and ethical concerns. Industry 4.0, on the other hand, revolutionised production through automation and networked systems. This development reflects a broader understanding of the value of human abilities and the need for a more responsible and balanced approach to technology growth. Sensitive data protection and improved cybersecurity measures become necessary when Industry 5.0 technologies and processes are integrated. The organisation must ensure robust cybersecurity procedures to safeguard employee and customer data privacy and defend against cyber-attacks [34–36].

## 1.5 Major Components of Industry 5.0

Industry 5.0 proposes a paradigm change that puts people at the centre of the industrial process, building upon the foundation of Industry 4.0. It acknowledges the limits of completely automated technologies and the unique skills of human labour. Cobots are a significant component of Industry 5.0. Industrial robots have continuously operated alone on the factory floor, fenced off in a secure area. Alternatively, cobots can be utilised in large-scale production to produce a semi-finished product, requiring human intervention at the last stage. They free up highly trained personnel to concentrate on reconfiguring and altering products by automating the tedious, hazardous, and repetitive elements of manufacturing [37–39]. Table 1.2 provides the significant components of Industry 5.0.

**Table 1.1** Main distinction of Industry 4.0 and Industry 5.0

| S. No. | Industry 4.0 | Industry 5.0 |
|---|---|---|
| 1 | Industry 4.0 provides strong emphasis on automation and minimising the need for human intervention in production processes. It uses cyber-physical systems, IoT, and AI to construct completely automated and self-optimising production settings | Industry 5.0 provides cooperation between humans and robots by returning the emphasis to human engagement. It blends the accuracy and efficiency of machines with the special talents of humans, such creativity and problem-solving |
| 2 | Industry 4.0 focusses primarily on efficiency and mass manufacturing. Large-scale production of standardised items is prioritised above some degree of customisation | Industry 5.0 emphasis on mass customisation and personalisation. Manufacturing companies can now produce goods that are customised to each client's tastes thanks to developments in AI and machine learning, which increases consumer happiness and creates new market prospects |
| 3 | It prioritises cost- and efficiency-cutting manufacturing process optimisation, perhaps putting less of a focus on sustainability | This gives environmental effect and sustainability a lot of attention. In order to reduce waste and carbon footprint, it encourages the adoption of circular economy ideas |
| 4 | This includes robots and autonomous systems that are made to function without much human supervision | It presents cobots, or collaborative robots, which augment human labourers' talents and foster a mutually beneficial partnership between humans and technology |
| 5 | Industry 4.0 uses AI and ML to optimise workflows and make judgements on its own, less emphasising moral issues | It stresses the value of reliable systems and moral AI. It promotes fair, transparent, and explicable AI algorithms to allay worries about bias and privacy while fostering human–machine trust |
| 6 | It focusses on automation and technological improvements | Industry 5.0 acknowledges the value of worker development and ongoing education. In order to provide workers with the skills they will need in future, it supports training programmes, collaborations between business and educational institutions, and lifelong learning initiatives |
| 7 | Industry 4.0 combines cloud computing, big data, IoT, and AI to build intelligent and networked production systems | Industry 5.0 expanding upon the technologies of Industry 4.0, this sector uses edge computing, blockchain technology, advanced robotics, human augmentation, and the Internet of behaviours (IoB) to improve human–machine collaboration and build more customised and sustainable production environments |
| 8 | Industry 4.0 focusses on efficiency and cost reduction by automating processes and analysing data in real time to optimise supply networks | Industry 5.0 makes use of blockchain and IoB technologies to improve supply chain traceability, transparency, and resilience. Its goal is to build supply chains that are more secure and flexible |

**Table 1.2** Components of Industry 5.0

| S. No. | Components | Description |
|---|---|---|
| 1 | Cooperative human–machine | Industry 5.0 strongly emphasises human–machine cooperation, whereas Industry 4.0 is more concerned with automation and minimising human participation. Collaborative robots, or cobots, complement human labourers by fusing their respective advantages. While robots offer precision, speed, and data processing capabilities, humans contribute creativity, flexibility, and problem-solving talents. As a result of this synergy, production processes become more adaptable and efficient |
| 2 | Sustainability | Sustainability is an essential component of Industry 5.0. Manufacturers are lessening their environmental impact, cutting waste, and implementing eco-friendly procedures. Combining sustainable materials, circular economy ideas, and renewable energy sources is becoming commonplace |
| 3 | Individualisation and tailoring | Industry 5.0 seeks to meet the increasing need for goods and services tailored to the individual. Manufacturers may now analyse consumer preferences and customise goods using AI and machine learning advancements to meet specific demands. This makes mass customisation possible and enables the large-scale creation of one-of-a-kind products. Bulk customisation replaces bulk production, increasing consumer satisfaction and creating new market opportunities |
| 4 | Workforce development | Due to the rapid evolution of technology, the workforce must continuously learn and acquire new skills. Industry 5.0 encourages programmes for lifelong learning to provide workers with the tools they need to succeed in a changing workplace |
| 5 | Lifelong learning | The workforce must be prepared for the future through collaborations between businesses and educational institutions, reskilling initiatives, and training programmes |
| 6 | Resiliance | Developing robust supply networks that can tolerate interruptions and adjust to changing circumstances is a major component of Industry 5.0 |

Manufacturing workers can profit significantly from Industry 5.0 technologies. They may focus on activities of more value since dull and repetitive chores are mechanised. Business productivity is significantly impacted by Industry 5.0. Integrating sophisticated technology enhances industrial operations' speed, precision, and efficiency. Humans and machines collaborate to maximise resources and boost output as a whole. Utilising the new processes and technologies unveiled in Industry 4.0, Industry 5.0 puts the human element back in the forefront and empowers individuals to generate more value. It requires a broad perspective on organisational performance, realising that social and human well-being are prerequisites for sustained corporate growth [40, 41].

## 1.6 Major Technologies of Industry 5.0

Various technologies of Industry 4.0 are used in Industry 5.0. However, some more technologies are helpful for Industry 5.0. The confluence of technologies that augment human skills and foster cooperation is what makes the shift to Industry 5.0 possible. Adopting Industry 5.0 has long-term advantages that are consistent with its fundamental principles. This entails increased resilience, significant energy savings, and better talent acquisition and retention. These benefits help businesses become more competitive and adaptable in changing markets and a dynamic environment. Reducing network traffic, enabling transactions, and protecting privacy are all made possible by intelligent manufacturing and intelligence [42–44]. The primary technologies of Industry 5.0 are discussed in Table 1.3.

Industry 5.0 also depends on technologies that enable people to manage resources efficiently and complete tasks, whether in the office or on the job. The asset management system should prioritise mobile devices and have an intuitive app to connect staff members continuously. This enables companies to effectively use software resources for data exchange across industrial sectors [3, 8, 9].

## 1.7 Application Areas of Industry 5.0

Industry 5.0 technology and ideas are used in several industries, changing how goods are created, produced, and distributed. It uses cognitive manufacturing systems, which produce more capable production processes by learning from experience and adapting to changing situations. Its main goal is to make human–machine interaction more intuitive and natural. For instance, speech and gesture recognition increase worker productivity and safety [10, 15, 21]. The primary application areas of Industry 5.0 are discussed in Table 1.4.

Numerous contemporary uses have been explored using Industry 4.0 technologies. These technologies have an excellent approach to troubleshooting unpredictable or harmful machinery to increase safety and efficiency. They continuously collect real-time data from IoT sensors, which AI-powered data analytics analyse [45–47].

## 1.8 Discussion

Industry 5.0 promotes the use of critical thinking to add value by personalising an item or service or incorporating human judgement into a decision-making process that was previously entirely automated. Newly identified synergies are sure to emerge when the inherent creativity of the human mind is combined with the potential of automation. Industrial automation, digitisation, 3D printing, quicker manufacturing, assembly line optimisation, increased accuracy, big data analytics, failure avoidance

**Table 1.3** Major technologies of Industry 5.0

| S. No. | Technologies | Description |
|---|---|---|
| 1 | Complex automation | Cobots are made to operate securely next to people. They can comprehend and react to human behaviours because they are outfitted with sensors and AI systems. Cobots help employees with repetitive motion, strength, and accuracy duties so people may concentrate on more intricate and creative work |
| 2 | Augmenting humans | Exoskeletons and wearable technology improve human potential and lessen physical strain. These tools help employees lift significant things, enhance their ergonomics, and operate more efficiently. Glasses with augmented reality technology offer real-time information and direction, improving decision-making and lowering mistakes |
| 3 | Edge computing | By bringing processing capacity closer to the point where data is generated, edge computing reduce latency and permits real-time processing. This is especially significant in production settings where prompt decision-making is essential. Edge devices provide local data analysis, which enables quicker reaction times and increased productivity |
| 4 | Blockchain technology | Blockchain technology makes supply chains more transparent and traceable. It ensures the authenticity and integrity of items by offering a safe, unchangeable record of transactions. Additionally, blockchain might make it easier for many industrial ecosystem players to collaborate and share secure data |
| 5 | Sophisticated machine learning and AI | More complex analysis and decision-making are possible as AI and machine learning algorithms advance. These technologies can foresee maintenance requirements, optimise manufacturing schedules, and instantly detect quality problems. Explainable AI ensures that human decision-making is transparent and understandable |

(continued)

**Table 1.3** (continued)

| S. No. | Technologies | Description |
|---|---|---|
| 6 | Behaviours on the Internet (IoB) | Expanding upon the idea of the IoT, the IoB uses data analysis from several sources to comprehend human behaviour. IoB has the potential to boost user experiences, streamline operations, and increase worker safety in Industry 5.0. Wearable technology may track employees' health and send notifications when they become tired or are in dangerous situations |
| 7 | Big data analytics | Industry 5.0 requires big data analytics because it improves the collaboration between human creativity and technologies. Analysing massive volumes of data from sensors, IoT devices, and other sources improves decision-making, allows real-time insights, and optimises production processes. Predictive maintenance, intelligent automation, and customised and sustainable production are all supported by this connection. By utilising big data, industries may increase productivity, decrease downtime, and satisfy customer needs for customised goods. Industry 5.0 innovation and productivity are driven by intelligent, adaptive systems made possible by big data analytics. These systems improve human–machine collaboration |

using cyber-physical systems, the IoB, cloud computing, and cognitive computing were all brought about by Industry 5.0. Predicting production efficiency based on ongoing activities is possible by utilising customised software, machine learning, industrial automation, and smart, networked devices. Flexibility is made possible by modifying procedures based on parameters to prevent losses. VR and AR have enormous potential in Industry 5.0 since they epitomise technology and human skills working together. Instead of providing typical classroom education, they can provide recently hired staff with real-world training, giving realistic scenarios. They also offer up countless possibilities by enabling the construction of intricate 3D virtual models for design and prototyping needs. Additionally, workers may receive instructions, coaching, and visual information overlays with AR-capable devices like smart glasses. This is especially useful for debugging complex jobs.

**Table 1.4** Application areas of Industry 5.0

| S. No. | Application areas | Description |
|---|---|---|
| 1 | Intelligent production | In smart manufacturing, production processes are optimised through the seamless collaboration of human workers and technology. Cobots help with material handling, assembly, and quality assurance, while AI algorithms examine data to find inefficiencies and bottlenecks. The result is a more responsive and agile production environment |
| 2 | Medical equipment and healthcare | The healthcare industry is greatly benefiting from Industry 5.0's contributions. Personalised medical equipment, such as implants and prosthetics, may be made utilising cutting-edge manufacturing processes to meet the specific demands of each patient. AI-powered telemedicine and diagnostics enable the continuous monitoring of patients and the conducting of remote consultations |
| 3 | Fashion and consumer products | The need for customised consumer goods is propelling innovation in the retail and fashion sectors. Customers can use internet platforms to create items, such as footwear and apparel, through mass customisation. 3D printing is one of the advanced production processes that swiftly and effectively realises these ideas |
| 4 | Automobile sector | The automotive sector is adopting Industry 5.0 to improve customer experiences and vehicle manufacturing. Collaborative robots assist human workers on assembly lines while enhancing productivity and output. AI systems examine sensor data to forecast maintenance requirements and to improve car performance |
| 5 | Logistics and supply chain | Industry 5.0 is revolutionising supply chain management and logistics operations by increasing resilience, traceability, and transparency. Blockchain technology lowers the possibility of counterfeiting and guarantees product authenticity. Real-time data analysis shortens lead times, improves customer happiness, and optimises inventory management |

## 1.9 Limitations of Industry 5.0

Industry 5.0 has great potential, but a few issues and concerns must also be considered. The significant limitations of Industry 5.0 are as follows:

***Change in Workforce***

The workforce must change in skills and capacities to incorporate sophisticated technology. Ensuring that employees are suitably trained and equipped for new positions is essential. Initiatives for reskilling, upskilling, and ongoing education programmes are crucial for facilitating this shift.

***Consequences for Ethics and Society***

The extensive usage of automation and AI raises moral and societal issues. Management is required to prevent algorithmic bias, data privacy, and employment displacement. To overcome these issues, regulations that support ethical behaviour and transparent and responsible AI systems are needed.

***Information security***

Cyberattack risk rises with the increased interconnectivity of production systems. Guaranteeing data and system security is crucial to avert interruptions and safeguard confidential data. Industry 5.0 requires strong cybersecurity measures like threat detection, access restrictions, and encryption.

***Frameworks for Regulation and Standardisation***

Technology is developing at a rate that frequently outpaces attempts at regulation and standardisation. Thorough frameworks that handle ethical, interoperability, and safety issues are essential. Establishing a conducive atmosphere for Industry 5.0 requires cooperation across standardisation organisations, governments, and the industry.

## 1.10 Future Research Scope

In future, Industry 5.0 will incorporate robotics, automation, and AI into manufacturing processes while giving workers a more active say in decision-making. It will also provide strong coordination between human and machine labourers, facilitating real-time information sharing.

With big data technology, IoT, and AI, businesses will enhance manufacturing strategies and processes, make better predictions, and provide a better work environment. They would have the flexibility to modify the company plan to meet customer demand instantly. Manufacturing in Industry 5.0 promises to use resources sensibly, according to the needs. Business models are adaptable when people and machines work together. As a result, excess output and waste can be minimised or eliminated.

The future belongs to the robot maker or the industrial robot. AI and cognitive computing are examples of cutting-edge technologies predicted to propel manufacturing to new heights of productivity and efficiency for both people and machines. Industry 5.0 will play a significant role in sustainability because its goal is to create renewable energy-powered systems. IoT devices, intelligent sensors, and specialised software all work together to monitor and anticipate potential issues before they happen. Only the machines most prone to malfunction will be halted so that changes can be made. The development of technologies like blockchain, IoT, and AI will soon determine how manufacturing progresses and how industries improve.

## 1.11 Conclusion

A human-centric industrial revolution is known as Industry 5.0. This is the subsequent stage of the fourth industrial revolution, which has unfolded over the past ten or so years. This evolution encourages mechanised efficiency and fosters human ingenuity, resulting in an ecosystem that allows humans and machines to collaborate more successfully and solve issues faster than in the past. Industry 5.0 represents the pinnacle of human–machine intelligence working with intelligent industrial equipment. Industry 5.0 prioritises sustainability as a novel approach to reduce production and maintenance costs in the long run instead of seeing it as merely another catchphrase. Industry 5.0 uses IoT sensors that continually monitor energy use, AI-powered analytical tools, and additive manufacturing to create a circular economy where all businesses reduce, reuse, and recycle. Industry 5.0 provides highly customised and customer-specific experiences. Man and machine can create jobs, improve worker safety and well-being, and expand flexibility to provide customers with personalised and interactive goods. Industry 5.0 is a potential stage of industrialisation in which robots driven by AI and cutting-edge technology interact alongside people to optimise workplace procedures. This evolution places a higher priority on a human-centric strategy that emphasises resilience and sustainability. In contrast to Industry 4.0, it aims to respect the constraints of production on our planet by using new technology to provide prosperity beyond job creation and economic development.

**Conflict of Interest** None.

## References

1. Maddikunta PK, Pham QV, Prabadevi B, Deepa N, Dev K, Gadekallu TR, Ruby R, Liyanage M. Industry 5.0: a survey on enabling technologies and potential applications. J Industr Inf Integr. 2022;26:100257.
2. Huang S, Wang B, Li X, Zheng P, Mourtzis D, Wang L. Industry 5.0 and Society 5.0—comparison, complementation and co-evolution. J Manuf Syst. 2022;64:424–8.
3. Bécue A, Gama J, Brito PQ. AI's effect on innovation capacity in the context of industry 5.0: a scoping review. Artif Intell Rev. 2024;57(8):215.
4. George AS, George AH, Baskar T. The evolution of smart factories: how industry 5.0 is revolutionizing manufacturing. Partners Univ Innov Res Publ. 2023;1(1):33–53.

5. Leng J, Sha W, Wang B, Zheng P, Zhuang C, Liu Q, Wuest T, Mourtzis D, Wang L. Industry 5.0: prospect and retrospect. J Manuf Syst. 2022;65:279–95.
6. Johri P, Singh JN, Sharma A, Rastogi D. Sustainability of coexistence of humans and machines: an evolution of industry 5.0 from industry 4.0. In: 2021 10th International conference on system modeling & advancement in research trends (SMART). IEEE; 2021. pp. 410–14.
7. Akundi A, Euresti D, Luna S, Ankobiah W, Lopes A, Edinbarough I. State of Industry 5.0—analysis and identification of current research trends. Appl Syst Innov. 2022;5(1):27.
8. Dmitrieva E, Balmiki V, Lakhanpal S, Lavanya G, Bhandari P. AI evolution in Industry 4.0 and Industry 5.0: an experimental comparative assessment. BIO Web Conf. 2024;86:01069.
9. Paschek D, Mocan A, Draghici A. Industry 5.0–the expected impact of next industrial revolution. In: Thriving on future education, industry, business and society; proceedings of the makelearn and TIIM international conference. ToKnowPress; 2019. pp. 125–32.
10. Jiménez-Partearroyo M, Medina-López A, Juárez-Varón D. Towards industry 5.0: evolving the product-process matrix in the new paradigm. J Technol Transfer. 2024;49(4):1496–531.
11. Milea A, Cioca LI. Work evolution and safety and health at work in Industry 4.0/Industry 5.0. MATEC Web Conf. 2024;389:00074.
12. Damodaran D, Ramathilagam A, SM US. Precision paradigm: AI-infused evolution of manufacturing in industry 5.0. In: Using real-time data and AI for thrust manufacturing. IGI Global; 2024. pp. 212–242.
13. Golovianko M, Terziyan V, Branytskyi V, Malyk D. Industry 4.0 vs. Industry 5.0: co-existence, transition, or a hybrid. Procedia Comp Sci. 2023;217:102–13.
14. Madsen DØ, Berg T, Di Nardo M. Bibliometric trends in industry 5.0 research: an updated overview. Appl Syst Innov. 2023;6(4):63.
15. Pereira R, dos Santos N. Neoindustrialization—reflections on a new paradigmatic approach for the industry: a scoping review on industry 5.0. Logistics. 2023;7(3):43.
16. Tallat R, Hawbani A, Wang X, Al-Dubai A, Zhao L, Liu Z, Min G, Zomaya AY, Alsamhi SH. Navigating industry 5.0: a survey of key enabling technologies, trends, challenges, and opportunities. IEEE Commun Surv Tut. 2023;26(2):1080–126.
17. Ghobakhloo M, Iranmanesh M, Mubarak MF, Mubarik M, Rejeb A, Nilashi M. Identifying industry 5.0 contributions to sustainable development: A strategy roadmap for delivering sustainability values. Sustain Prod Consump. 2022;33:716–37.
18. Majerník M, Daneshjo N, Malega P, Drábik P, Barilová B. Sustainable development of the intelligent industry from Industry 4.0 to Industry 5.0. Adv Sci Technol Res J. 2022;16(2):12–8.
19. Tyagi AK, Lakshmi Priya R, Mishra AK, Balamurugan G. Industry 5.0: potentials, issues, opportunities, and challenges for society 5.0. In: Privacy preservation of genomic and medical data. 2023. pp. 409–32.
20. Babkin A, Shkarupeta E, Kabasheva I, Rudaleva I, Vicentiy A. A framework for digital development of industrial systems in the strategic drift to Industry 5.0. Int J Technol. 2022;13(7).
21. Li X, Nassehi A, Wang B, Hu SJ, Epureanu BI. Human-centric manufacturing for human-system coevolution in Industry 5.0. CIRP Annals. 2023;72(1):393–6.
22. Zeb S, Mahmood A, Khowaja SA, Dev K, Hassan SA, Gidlund M, Bellavista P. Towards defining industry 5.0 vision with intelligent and softwarized wireless network architectures and services: a survey. J Netw Comp Appl. 2024;223:103796.
23. Slavic D, Marjanovic U, Medic N, Simeunovic N, Rakic S. The evaluation of Industry 5.0 concepts: social network analysis approach. Appl Sci. 2024;14(3):1291.
24. Rožanec JM, Novalija I, Zajec P, Kenda K, Tavakoli Ghinani H, Suh S, Veliou E, Papamartzivanos D, Giannetsos T, Menesidou SA, Alonso R. Human-centric artificial intelligence architecture for industry 5.0 applications. Int J Prod Res. 2023;61(20):6847–72.
25. Ciucu-Durnoi AN, Delcea C, Stănescu A, Teodorescu CA, Vargas VM. Beyond Industry 4.0: tracing the path to Industry 5.0 through bibliometric analysis. Sustainability. 2024;16(12):5251.
26. Pizoń J, Gola A. Human–machine relationship—perspective and future roadmap for industry 5.0 solutions. Machines. 2023;11(2):203.
27. Ordieres-Meré J, Gutierrez M, Villalba-Díez J. Toward the industry 5.0 paradigm: Increasing value creation through the robust integration of humans and machines. Comp Industry. 2023;150:103947.

28. Zizic MC, Mladineo M, Gjeldum N, Celent L. From industry 4.0 towards industry 5.0: A review and analysis of paradigm shift for the people, organization and technology. Energies. 2022;15(14):5221.
29. Mathur A, Dabas A, Sharma N. Evolution from industry 1.0 to industry 5.0. In: 2022 4th International conference on advances in computing, communication control and networking (ICAC3N). IEEE; 2022. pp. 1390–4.
30. Möller DP, Vakilzadian H, Haas RE. From Industry 4.0 towards Industry 5.0. In: 2022 IEEE international conference on electro information technology (eIT). IEEE; 2022. pp. 61–8.
31. Lagorio A, Cimini C, Romero D. A reflective framework for understanding workforce evolutionary pathways in industry 5.0. In: IFIP international conference on advances in production management systems. Cham: Springer Nature Switzerland; 2023. pp. 501–12.
32. Chander B, Pal S, De D, Buyya R. Artificial intelligence-based internet of things for industry 5.0. In: Artificial intelligence-based internet of things systems. 2022. pp. 3–45.
33. Alvarez-Aros EL, Bernal-Torres CA. Technological competitiveness and emerging technologies in industry 4.0 and industry 5.0. Anais da Academia Brasileira de Ciências. 2021;93:e20191290.
34. Yavari F, Pilevari N. Industry revolutions development from Industry 1.0 to Industry 5.0 in manufacturing. J Industr Strat Manage. 2020;5(2):44–63.
35. Aslam F, Aimin W, Li M, Ur Rehman K. Innovation in the era of IoT and industry 5.0: absolute innovation management (AIM) framework. Information. 2020;11(2):124.
36. Longo F, Padovano A, Umbrello S. Value-oriented and ethical technology engineering in industry 5.0: a human-centric perspective for the design of the factory of the future. Appl Sci. 2020;10(12):4182.
37. Alexa L, Pîslaru M, Avasilcăi S. From Industry 4.0 to Industry 5.0—an overview of European Union enterprises. In: Sustainability and innovation in manufacturing enterprises: indicators, models and assessment for industry 5.0. 2022. pp. 221–31.
38. De Souza RO, Ferenhof HA, Forcellini FA. Industry 4.0 and Industry 5.0 from the Lean perspective. Int J Manag Knowl Learn 2022;11:145–55.
39. Jefroy N, Azarian M, Yu H. Moving from Industry 4.0 to Industry 5.0: what are the implications for smart logistics? Logistics. 2022;6(2):26.
40. Mourtzis D. The Metaverse in industry 5.0: A human-centric approach towards personalized value creation. Encyclopedia. 2023;3(3):1105–20.
41. Grabowska S, Saniuk S, Gajdzik B. Industry 5.0: improving humanization and sustainability of Industry 4.0. Scientometrics. 2022;127(6):3117–44.
42. Xu X, Lu Y, Vogel-Heuser B, Wang L. Industry 4.0 and Industry 5.0—inception, conception and perception. J Manuf Syst. 2021;61:530–5.
43. Adel A. Future of industry 5.0 in society: human-centric solutions, challenges and prospective research areas. J Cloud Comput. 2022;11(1):40.
44. Paschek D, Luminosu CT, Ocakci E. Industry 5.0 challenges and perspectives for manufacturing systems in the society 5.0. In: Sustainability and innovation in manufacturing enterprises: indicators, models and assessment for industry 5.0. 2022. pp. 17–63.
45. Skobelev PO, Borovik SY. On the way from Industry 4.0 to Industry 5.0: from digital manufacturing to digital society. Industry 4.0. 2017;2(6):307–11.
46. Doyle-Kent M, Kopacek P. Industry 5.0: is the manufacturing industry on the cusp of a new revolution? In: International conference on production research. Cham: Springer International Publishing; 2019. pp. 432–41.
47. Bryndin E. Formation and management of Industry 5.0 by systems with artificial intelligence and technological singularity. Am J Mech Ind Eng. 2020;5(2):24–30.

# Chapter 2
# Introduction to Photonics and Optoelectronics in Industry 5.0

**Amit Das, Preeti Goyal, Swati Sharma, and R. S. Gupta**

**Abstract** This chapter discusses the pivotal role that photonics and optoelectronics play in advancing modern industry. The integration of human expertise with innovative technologies in Industry 5.0 underscores the significance of photonics and optoelectronics in advancing more efficient and intelligent manufacturing processes. As Industry 5.0 focuses on integrating human intelligence with cutting-edge technology, these fields are instrumental in creating smarter, more efficient manufacturing processes. Photonics and optoelectronics enable precise communication, sensing, and automation, driving innovation in areas such as robotics, smart factories, and sustainable production. Their contributions extend to improving energy efficiency, reducing material waste, and fostering environmentally friendly manufacturing practices. As Industry 5.0 emerges, focusing on the synergistic relationship between human creativity and advanced technologies, photonics and optoelectronics are crucial in fostering smarter and more efficient manufacturing methods. This chapter provides an introduction to thcsc technologies and their applications, highlighting their significance in shaping the future of industrial development and their potential to redefine industry standards in the years ahead.

**Keywords** Automation · Human–machine collaboration · Industry 4.0 · Industry 5.0 · Intelligent systems · Photonics · Optoelectronics · Sensors · Smart factory

---

All the authors have contributed equally in this work.

---

A. Das (✉)
Department of Electronics and Communication Engineering, GLA University, Mathura, Uttar Pradesh, India
e-mail: amitofficial7492@gmail.com

P. Goyal (✉) · S. Sharma (✉) · R. S. Gupta
Department of Electronics and Communication Engineering, Maharaja Agrasen Institute of Technology, New Delhi, India
e-mail: preetigoyal.s@gmail.com

S. Sharma
e-mail: swati1986sharma@gmail.com



V. Bhardwaj et al. (eds.), *Photonics and Optoelectronics in Industry 5.0*, Advances in Optics and Optoelectronics, https://doi.org/10.1007/978-981-96-7594-4_2

## 2.1 Introduction

### *2.1.1 Overview of Photonics and Optoelectronics*

Optoelectronics or photonics or the use of light is a branch that combines ideas from optics, electronics, physics, and material science in order to enhance the growth in several sectors. This field deals with the generation, processing, and measurement of light, including ordinary, ultraviolet, and infrared beams of light. Photonics is crucial in the different technological sectors such as communication technology, medical technology, engineering technology, and electronics technology. Optical fibers are used in telecommunication to enable the transfer of large volumes of data over long spans with negligible attenuation. In medicine, photonics has led to advances such as laser surgery, optical coherence, tomography, and advanced imaging. Laser-based cutting, welding, and material processing are added advantages to the manufacturing processes due to its accuracy and speed. Furthermore, photonic components improve the operation and reliability of personal gadgets including cameras, smart phones, and displays [1]. In general, photonics is one of the major enablers of technological evolution and a constantly evolving process helping to enhance our current reality (modernization). Transistors are effectively utilized for different applications in the current era [2–10].

Optoelectronics is the key subdivision of photonics and deals with the dynamic interaction between light and electronics. This field is at the root of the effortless conversion of electrical signals into light or vice versa leading to development in various fields. Its core consists of technologies which change the existing markets, for example, light emitting diodes (LEDs) transforming the lighting and display industry through power efficient illumination. Laser diodes are another innovation which produces focused light beams in nearly all processes (vital applications) including surgeries, data transfers among others. Photodetector is the vision of the current technology where light is converted to an electrical signal for image and sensing applications. At the same time, solar cells harness energy from the sun, initiating the crucial shift toward clean energy, which is highly needed in today's world. These optoelectronic marvels are the building blocks of contemporary telecommunication industry and improve the data rates and bandwidth of international communication web. In healthcare, they enable advanced diagnostic tools and treatments, pushing the boundaries of medical science. The consumer electronics industry relies heavily on optoelectronics to create smarter and more responsive devices that enrich our daily lives. By bridging the realms of electronics and optics, optoelectronics not only enhances existing technologies but also paves the way for groundbreaking innovations, continuously reshaping our technological landscape and propelling us toward a more connected and efficient future [11]. Figure 2.1 shows the typical synoptic overview of photonics and optoelectronics.

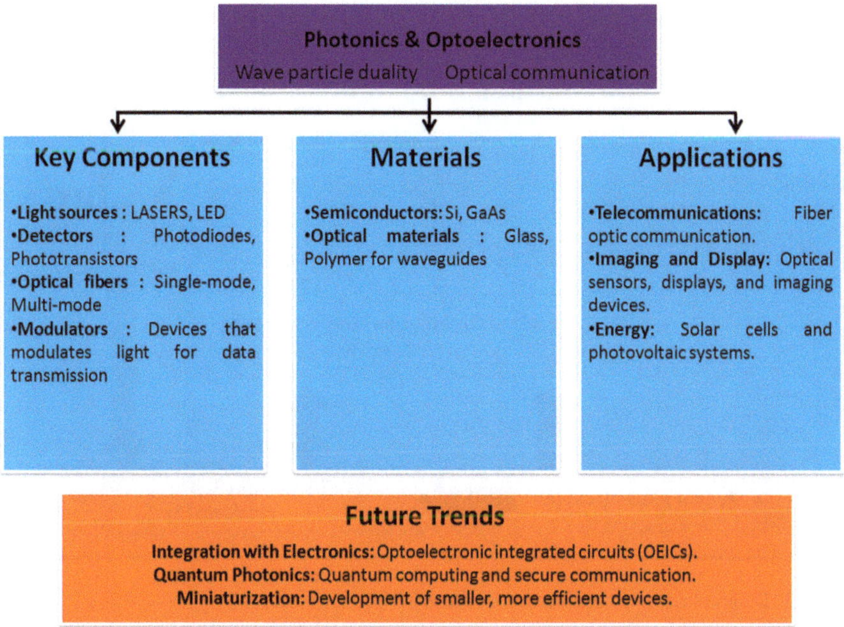


**Fig. 2.1** Synoptic overview of photonics and optoelectronics

### *2.1.2 Overview of Industry 5.0*

Industry 5.0 marks a paradigm shift in industrial evolution, building upon the foundations laid by Industry 4.0. This new era re-imagines and redefines the relationship between humans and technology, emphasizing synergy rather than replacement. At its core, Industry 5.0 champions a human-centric approach, recognizing that while machines excel at precision and efficiency, human creativity and intuition remain irreplaceable [12]. This philosophy fosters a collaborative environment where advanced technologies augment human capabilities, leading to a more innovative problem-solving and decision-making process. A hallmark of Industry 5.0 is its focus on personalization at scale. As consumer demand for unique products grows, this new industrial model harnesses cutting-edge technologies to enable mass customization. This approach efficiently integrates the efficiency of large-scale production with the flexibility to tailor products to individual preferences, revolutionizing manufacturing paradigms. Sustainability forms the third pillar of Industry 5.0, reflecting a growing awareness of environmental responsibilities. This model integrates eco-friendly practices into every aspect of industrial processes, from energy-efficient operations to waste reduction and responsible resource management [12]. Concluding it, the key features of Industry 5.0 include

(i) Human-centric approach.
(ii) Personalization.

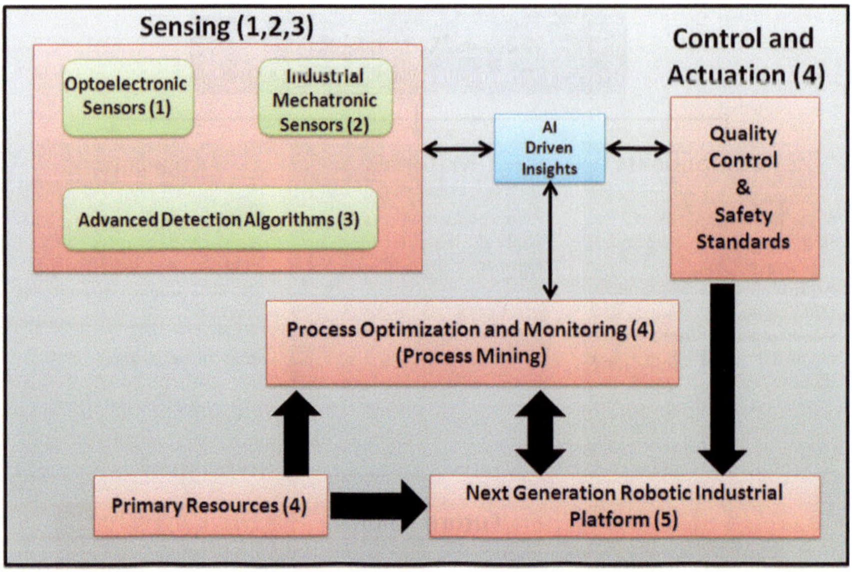


**Fig. 2.2** Diagrammatic representation of Industry 5.0 framework [13]

(iii) Sustainability.

By prioritizing sustainability, Industry 5.0 not only addresses environmental concerns but also aligns industrial growth with long-term ecological balance, paving the way for a more responsible and sustainable future of manufacturing.

Figure 2.2 illustrates a block diagram sketching a basic Industry 5.0 framework (given (idea) by Massaro [13]), which will be discussed in this chapter. The diagram, deduced from the analysis of the state of the art and from emergent topics in research, consists of the following blocks or grouped modules:

- A sensing module composed of innovative optoelectronic sensors (block 1), industrial mechatronic sensors (block 2), and detection algorithms (block 3).
- Control and actuation facilities (*block 4*) which address production, quality, safety, process management, raw materials management, and process mining (PM).
- An AI engine managing production line actions (*blocks 1, 2, 3, 4, 5*).
- An Advanced robotic industrial platform (*block 5*).

## *2.1.3 Differences Between Industry 4.0 and Industry 5.0*

While both Industry 4.0 and Industry 5.0 leverage advanced technologies to revolutionize manufacturing and industrial processes, their core philosophies, focus areas, and implementation strategies differ significantly:-

***Industry 4.0***

(a) **Automation-centric**: Industry 4.0 places a strong emphasis on automating processes to increase efficiency and reduce human intervention. The goal is often to minimize human involvement in routine tasks.
(b) **Connectivity and data**: It focuses heavily on the Internet of things (IoT), creating interconnected systems that generate and analyze vast amounts of data in real-time.
(c) **Artificial intelligence**: AI and machine learning are used primarily for process optimization, predictive maintenance, and autonomous decision-making.
(d) **Efficiency-driven**: The primary aim is to enhance productivity and operational efficiency through digital transformation.
(e) **Cyber-physical systems**: Industry 4.0 introduces the concept of cyber-physical systems, where physical processes are monitored and controlled by computer-based algorithms.
(f) **Big data analytics**: Utilizes advanced analytics to derive insights from large datasets, enabling data-driven decision-making.
(g) **Digital twins**: Creates virtual representations of physical assets to simulate and optimize processes.

***Industry 5.0***

(a) **Human-centric**: While still leveraging automation, Industry 5.0 reintroduces human involvement, focusing on collaboration between humans and machines.
(b) **Cognitive computing**: Goes beyond mere connectivity to implement cognitive systems that can understand, learn, and interact more naturally with humans.
(c) **Augmented intelligence**: Rather than replacing humans, AI is used to augment human capabilities, enhancing decision-making and creativity.
(d) **Sustainability focus**: Incorporates environmental sustainability as a core principle, aiming to reduce waste and energy consumption.
(e) **Mass customization**: Utilizes advanced technologies to enable personalized production at scale, catering to individual consumer preferences.
(f) **Ethical considerations**: Places greater emphasis on the ethical implications of technology, including job displacement and data privacy.
(g) **Biologically inspired systems**: Incorporates bio-mimicry and bio-inspired technologies to create more adaptive and resilient industrial systems.
(h) **Holistic approach**: Considers the broader societal impact of industrial processes, including worker well-being and community development.
(i) **Circular economy integration**: Emphasizes the principles of a circular economy, focusing on resource recovery and regenerative design.
(j) **Human–robot collaboration**: Develops advanced robotics systems designed to work alongside humans safely and effectively, rather than replace them.

While Industry 4.0 concentrates on establishing intelligent factories by adopting smart technologies and the process of interoperability, Industry 5.0 stresses the development of wise factories by improving the application of smart technology, human well-being, and sustainable environment. It signified a transition from an approach which was dominated by the use of machines toward a comprehensive person-oriented and eco-friendly integrated industrial model. While both Industry 4.0 and Industry 5.0 involve advanced technologies, their focus areas differ largely:-

**Industry 4.0**: Stresses on the idea of automation, connectivity and the usage of real-time data. These include Internet of things, artificial intelligence, and big data.

**Industry 5.0**: Emerges from Industry 4.0 and puts an overlay of teamwork between humans and smart machines integrated with sustainability. It aims at incorporating the input of human imagination and skill in the rigid structures developed by Industry.

### *2.1.4 Emphasis on Human–Machine Collaboration, Customization, and Sustainability*

Industry 5.0 is characterized by an increased cooperation between humans and robots; thus, each contributes to the performance of the other. Special effort is devoted to decision-making, thinking, innovations and creativity as machines handle repetitive and accurate work, hence increasing the overall efficiency [14]. Additionally, the shift toward customization is emphasized as a key characteristic, with modern trends like additive manufacturing, including 3D printing and flexible manufacturing systems, enabling the production of individualized components with minimal changeover and reduced costs. Sustainability is also embedded in Industry 5.0 in terms of environmentally sustainable means through utilization of renewable products, green production processes, circular economy principles to reduce the use of resources and maximize their productivity [12]. Photonics and optoelectronics are at the base of up-to-date manufacturing and smart factories providing increased accuracy, speed, and variability to industrial processes. Laser application is considered as highly significant and can be used in understanding manufacturing that involves laser cutting, laser welding, laser marking, and laser sintering or metal 3D printing. They are highly accurate, versatile, and irreplaceable for manufacturing of complex and highly accurate (precision) parts from various materials. Another important sensor technology is optical sensors, which supply contactless, real-time information about characteristics including temperature, pressure, or composition of a material [15]. In smart factories, they use these sensors to monitor and control the quality of products and reduce on wastage, since the production is on real-time basis.

Optoelectronic devices offer significant importance in improving the human–machine interface by making the interaction between human and machines easier. Mobile virtual reality (VR) and augmented reality (AR) systems are based on optoelectronics components including the smart-display, camera, and sensor. These systems are used for activities such as training, maintenance, and tele-support, which may enhance safety and efficiency of workplaces. Furthermore, gesture recognition and more developed interfaces of man/machine rely on the optoelectronic sensors for touchless operations; in this way, operators can control machines and systems and use them most naturally. They not only help reduce the probability of making errors but also improve clients' satisfaction level.

Photons are at the base of most contemporary forms of communication technology acting as the framework for high-speed transmission of data and solid industrial applications. Fiber optics is instrumental in this regard and takes advantage of light to transmit data at huge and very-fast speeds over long distances and with relatively little signal attenuation. It guarantees flexible and effective data transmission in industries and helps in actual time tracking and controlling of tools and machinery. However, photonic integrated circuits (PICs) integrate several photonic functional blocks in a single chip making communication systems compact and efficient. A PIC is used in data centers, telecommunications system, and sensor networks and plays a critical role in the data processing and connection, which are important in the current industrial world.

Photonics and optoelectronics are critical components in promoting the concept of cleaner manufacturing and renewable energy systems. LED and OLED (organic LED) lighting improves virtually all aspects of lighting by providing significantly greater efficiency and longer product lifetime than incumbent technologies. The use of these technologies in industries leads to cutting down of both the energy and costs of maintaining the industries. Secondly, photovoltaic and the other solar energy solutions encompass the application of the photonic technologies in generation of effectual solar cells. These photovoltaic systems can turn sunlight into electricity, giving industries an opportunity to reduce on the use of fossil fuels (utilizing cleaner and sustainable energy).

## 2.2 Key Technologies and Innovations

### *2.2.1 Laser Technology*

Laser is a basic element of the photonic systems, which is a key component in different spheres of industry as far as their precision, power, and flexibility are very significant and crucial. Among all the kinds of lasers, $CO_2$ lasers are gas lasers that work at the infrared spectrum, widely applied for cutting, welding, and engraving materials such as metals, plastics, ceramics. Fiber lasers, which use optically active fibers doped with rare-earth elements, offer significant advantages in efficiency and size, along with high beam quality, making them ideal for cutting, welding, and marking applications [16]. The diode laser also termed as the semiconductor laser is another type of laser renowned for high efficiency and dependability in applications such as bar-code scanning, medical uses, and processing of material.

In terms of application, laser cutting is well known for being able to make clean, accurate, and precise cuts in a variety of materials of metal, plastic, and composites and is widely used in the automotive, aerospace, and electronics industries. Laser

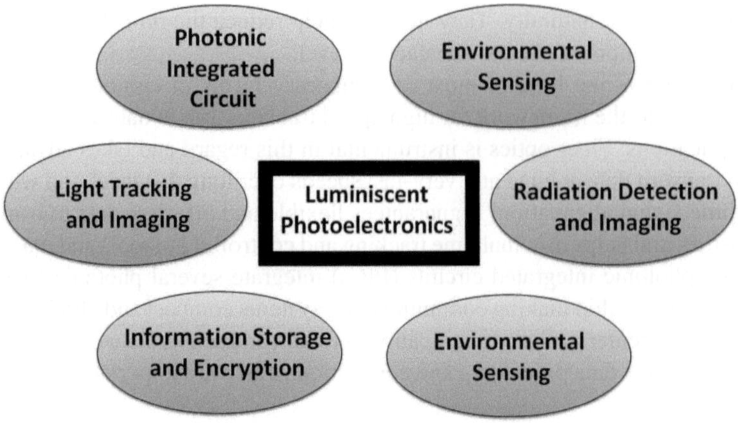


**Fig. 2.3** Key photonic application areas in luminescent optoelectronics

welding provides really strong and accurate joint for which latter is very important while manufacturing complex assemblies and high-precision components; laser welding is widely used in automotive, electronics, and medical industries. Additionally, laser-based additive manufacturing, or 3D printing, enables the creation of complex and customized parts layer-by-layer, transforming prototyping and production across sectors like aerospace, healthcare, and consumer goods. Figure 2.3 shows the key photonic application areas of luminescent optoelectronics that leverage the versatile integration of luminescent materials and optoelectronic platforms [5, 17].

### *2.2.2 Optical Sensors and Imaging Systems*

Optical sensors and imaging systems are essential to modern industrial processes, offering precise and real-time data for effective monitoring, control, and quality assurance. Among the different types of sensors, Light Detection and Ranging (LIDAR) utilizes laser pulses to measure distances and create detailed 3D maps of environments, with applications spanning from autonomous vehicles to industrial automation, and environmental monitoring. Cameras, including those using visible light, infrared, and ultraviolet, are integral to machine vision systems, facilitating tasks such as inspection, measurement, and guidance in manufacturing. Photodiodes and phototransistors detect light and convert it into electrical signals, finding uses in safety systems, environmental monitoring, and communication devices. In terms of applications, machine vision systems, powered by optical sensors and cameras, are critical for inspecting products, identifying defects, and ensuring quality control in industries like electronics, pharmaceuticals, and automotive manufacturing. Environmental monitoring also benefits from optical sensors, which track parameters such as air quality, water purity, and chemical concentrations, aiding industries in

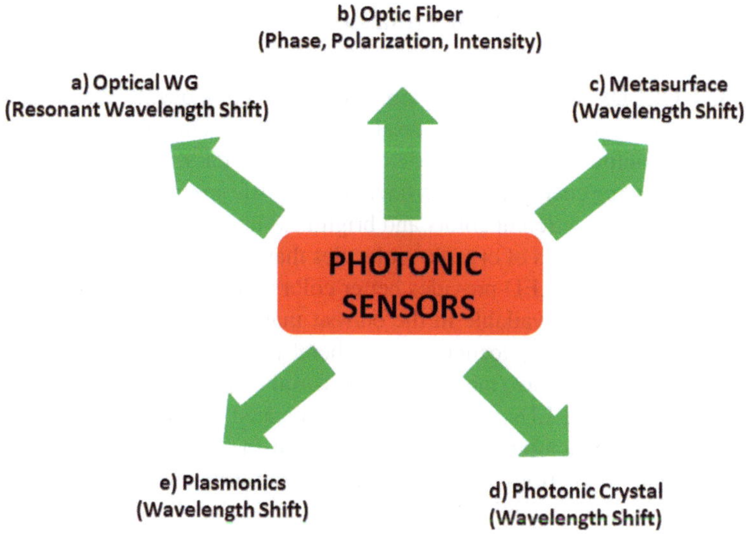


**Fig. 2.4** Photonic sensing technologies based on **a** optical waveguide, **b** optic fiber, **c** meta surface, **d** photonic crystal, and **e** plasmonic [18]

regulatory compliance and maintaining safe working conditions. Figure 2.4 displays the key sensing technologies in photonics such as optic fiber, optical waveguide, plasmonic, etc. [18].

### *2.2.3 Fiber Optics*

Fiber optics technology is fundamental for achieving high-speed communication and reliable data transmission in industrial environments. Fiber optic cables are composed of thin strands of glass or plastic that transmits light signals over long distances with minimal signal loss (attenuation). This capability allows them to handle large volumes of data at high speeds efficiently. Examples of the uses for fiber optic cables include industrial communication, including high-speed Internet connections. This technology helps to communicate data in real-time between machines, sensors, and controlling systems further promoting operation's effectiveness and dependability. Furthermore, fiber optic sensors are used to measure structural health, temperature, pressure, and other parameters in real-time operating conditions [19]. These sensors are mostly applicable in aerospace engineering, civil engineering, and production of energy, since they provide constant and reliable data.

### 2.2.4 *Light Emitting Diodes and Organic LED*

LEDs and OLEDs are the critical optoelectronic components that find use in numerous commercial and industrial products. LEDs are semiconductor devices that give out light (output) once the electric current (input) passes through it. These are recognized for being energy-efficient, having a long-life span and also the possibility of being produced in different colors and brightness levels [20]. In contrast, OLEDs are made of new materials of organic molecules that glow once an electric current is passed through them. OLED provides better colors, flexibility, and slimmer design than the general LEDs available in the current market. Some of the uses of these technologies are in display technologies, whereby LEDs and OLEDs are used in the television, smart phones, tablets, and industrial control panels. Of all the types of modern displays, OLEDs are especially appreciated for their contrast ratios and viewing angles. Application of LEDs is common in industrial, commercial, and residential lighting solutions because of their higher efficiency and long-life span. OLEDs are also finding application in various lighting applications in the form of flexible and stylish products.

## 2.3 Practical Examples and Applications

***Scenario 1****: Smart Factory Implementation*

Smart factory efficiently combines photonic and optoelectronic systems to enhance automation and increase production speed and personalization. For instance, many biggest automobile manufacturers in the world has adopted a smart plant concept which employs laser technology for cutting and joining of auto components [21]. This approach enhances high accuracy and quality in production. The uses of optical sensors are for the detection of errors and enhancing production quality, inline in order to minimize defects and therefore waste. Consequently, the factory enjoys higher productivity, less time-off, and the opportunity to bring out models tailored to the current market [22]. The use of the real-time monitoring system ensures that quality of the final products is enhanced and material usage is made more effective in a manner that makes the system more sustainable and responsive to the manufacturing process.

***Scenario 2****: Enhanced Human–Machine Interaction*

The use of augmented and virtual reality in interactions between man and machine is revolutionizing training, maintenance, and operation in many ways. For instance, many manufacturing firms have integrated AR glasses with optical sensors and the glasses are used to provide directions to the assembly line personnel [23]. This basically superimposes instructions and data on the worker's vision space to enable precise and timely work accomplishment. Also, VR systems are used for training so that the employees can exercise in front of virtual equipment before they work with real one. Advantages of this implementation are immediate feedback and direction (guidance) to the workers, hence decreasing the rate of making mistakes (errors) and increasing efficiency. VR training programs enhance safety and preparedness by allowing employees to train virtually, ensuring they are as prepared as possible when actual training occurs [14]. Collectively, utilizing AR and VR improves the usability and raises the security of working environments.

***Scenario 3**: Sustainable Manufacturing*

Photonics and optoelectronics are pivotal in advancing sustainable manufacturing practices by minimizing energy consumption and reducing environmental impact. For instance, few solar panel producers have successfully applied laser solutions in order to increase the performance of the photovoltaic cells, as well as the amount of energy produced. Also, proper lighting systems such as LED lighting system are used in the facility to reduce energy use by a large margin. The advantages of these practices include increased efficiency of energy used in production processes, hence decreasing the company's environmental impact. Furthermore, the integration of energy-efficient lighting and the latest technologies to optimize renewable energy components reinforces the company's commitment to sustainable production and enhances its environmental responsibility [24].

## 2.4 Photonics and Optoelectronics in Specific Industries

Optoelectronics and photonics are rapidly becoming important in automotive applications due to their ability to improve safety, performance, and even manufacturing processes. LIDAR systems use laser pulse scanners to create detailed representations of the environment, enabling intelligent transport systems to operate safely and efficiently. In the same respect, optical sensors are of great importance in the topic of obstacle detection as well as monitoring the vehicle environment, significantly enhancing safety in automobiles by supporting advanced driver assistance systems (ADAS). Laser-based manufacturing is used for cutting, welding, and marking of automotive products, which increases the quality of manufacturing and decreases the usage of material. This technology also facilitates the manufacture of low density and compound shapes for light weight structures and improved fuel efficiency as well as vehicle design.

Photonics and optoelectronics are central to healthcare, enhancing the diagnosis and treatment of diseases, as well as advancing research. Methods like the optical coherence tomography (OCT) are used in getting cross-sectional images of body tissues; it can be used in diagnosis of eye diseases, cardiovascular diseases as well as cancer. Fluorescence and Raman spectroscopy have diagnostic functions to deliver diagnostic data without invasiveness, enabling early diagnostics and disease monitoring. Wearable health monitors and biosensors are equipped with optoelectronic sensors to continuously track vital signs, including heart rate, oxygen levels, and glucose concentrations [25–27]. These devices offer real-time health data to both patients and healthcare providers, while biosensors detect specific biomarkers in bodily fluids, enabling rapid and accurate diagnostics at the point of care [7, 28, 29].

Photonics is fundamental to modern telecommunications, enabling high-speed and high-capacity data transmission. Fiber optic cables are the backbone of 5G networks, supporting the high data rates and low latency required for next-generation communication. Photonic components such as wavelength-division multiplexing (WDM) and coherent optical systems enhance the capacity and efficiency of data transmission. In fiber optic communication, fiber optic networks handle the enormous volumes of data generated and processed daily. Photonics ensures fast and reliable data transfer within and between data centers. Moreover, photonic integrated circuits reduce power consumption and increase the density of communication links, thereby improving the overall efficiency of data centers.

## 2.5 Emerging Applications and Innovations

**Quantum photonics**: Quantum photonics is known as the application of elements from quantum mechanics to develop novelty technologies with higher performance. This field focuses on the control of single photons and entangled photon-pairs for use in communication, computation and sensing. For instance, quantum key distribution (QKD) uses the quantum nature of light particles called photons to make secure communication and at the same time monitor any act of wiretapping. Also, quantum computing makes the use of quantum bits or qubits in order to solve complicated computation problems much more than the capability of conventional computing systems. The quantum sensors give highly accurate measurements of certain physical values like magnetic field intensity, gravitational forces, and time and can be used in navigation systems, geological surveys, and studies of physical phenomenon at a basic level [30].

**Integrated photonics**: Integrated photonics involves packaging of more than one photonic function on the same chip, hence miniaturization and enhancement of performance [31]. A PIC incorporates laser, modulators, detectors, wave guides, etc. on a single chip and thus has beneficial effects on size, power consumption, and cost. PICs are applied to the high-speed data communication, biosensors, and quantum information technology. They lead to the design of high-speed optical transceivers for data centers and telecommunication networks and portable and compact sensor systems for environmental, healthcare, and industrial applications.

**Nanophotonic**: Nanophotonic studies optical behavior of objects at nanoscale and thus results into new optical effects and its related uses. Instruments typical of this field interacts with light at scales below or at the same order as the wavelength of the light used, providing a high level of control over its transmission, absorption, and emission. Plasmonic, a subfield of nanophotonic, utilizes the interaction of light with metal nanostructures to achieve intense light confinement and enhancement [32]. Applications include super-resolution imaging techniques like stimulated emission

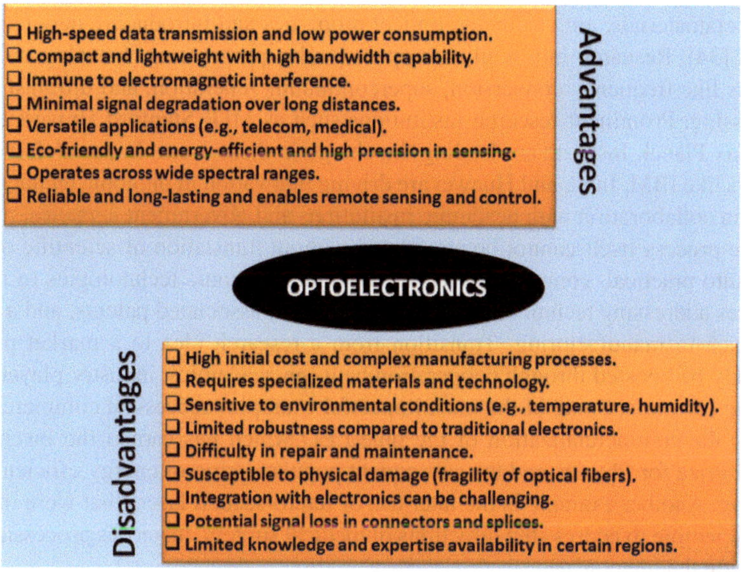


**Fig. 2.5** Advantages and disadvantages of optoelectronics

depletion (STED) microscopy, which provides detailed views at the nanoscale for advancements in biological and materials research. Nanophotonic sensors offer high sensitivity and specificity for detecting minute changes in environmental conditions or biological samples [28, 33]. Additionally, metamaterials made from nanostructures enable innovative light manipulation capabilities such as negative refraction and cloaking, with potential uses in imaging, communication, and security.

Without a doubt, optoelectronics is a vital technology that has enabled practically a seamless function of the information industry, and it has benefitted numerous sectors that are involved, such as medical, communications, aerospace, and defense. However, like any technology, there are key advantages and disadvantages (shown in tabular form in Fig. 2.5) of optoelectronics that need to be considered.

## 2.6 Photonics and Optoelectronics in Research and Development

Ongoing research in photonics and optoelectronics is at the forefront of technological advancement, exploring innovative applications and pushing the boundaries of the field. Current research includes the development of new laser sources with enhanced power, efficiency, and wavelength versatility, tailored for both industrial and medical uses. Advances in photonic materials, such as two-dimensional materials and metamaterials, are enhancing optical properties and introducing new functionalities [34]. Research into nonlinear-optics is also progressing, focusing on applications like frequency conversion, supercontinuum generation, and ultrafast signal processing. Prominent research institutions such as MIT, Stanford University, and the Max Planck Institute are leading the charge in these areas. In parallel, industry leaders like IBM, Intel, and Huawei are driving innovation in photonic technologies, often in collaboration with academic institutions and government agencies.

The process itself cannot be completed without translation of scientific discoveries into practical, commercial usages. Bringing photonic technologies to market requires addressing technical barriers, protection of associated patents, and a sound approach to manufacturing. Transition from a research idea to a market product can only be boosted through cooperation between academics, industry players, and leading investors. Some of the most impactful examples successful commercialization is the commercialization of the blue-LED which has earned the inventors a Nobel Prize for Physics and the subsequent popularization of energy-efficient LED lighting. Another example is the later use of femto-second lasers that were initially opto-scientific to address the directions of medical surgery, materials processing and precision manufacturing.

## 2.7 Standards and Regulations

**Industry standards**: The need to make photonic and optoelectronic technologies compatible, inter-operable, and of high quality makes standardization an essential step. Standards by the International Electrotechnical Commission (IEC) and the Institute of Electrical and Electronics Engineers (IEEE) are used to set benchmarks for the manufacturing of photonic devices and systems [17]. These standards cover many areas: performance characteristics, safety norms, and evaluation procedures among others. If norms are not clearly defined and sustained, there is vulnerable risk for electronics manufacturing and communications to be unreliable in terms of quality and capability to perform in various applications of photonic technologies.

**Regulatory considerations**: There is no doubt that regulations have a central function of protecting people's safety and the environment when using photonic and optoelectronic technologies. Laser and optical radiation standards apply to the utilization in order to minimize risks that may affect employees and the society. Environmental regulations also cover the issue of waste disposal and recycling of photonic components like LEDs and solar cells. In addition to complying with standards set by the International Electrotechnical Commission (IEC) and the Institute of Electrical and Electronics Engineers (IEEE) for photonic devices and systems, manufacturers in the USA must also adhere to regulatory specifications from agencies such as the Food and Drug Administration (FDA) for medical devices and the Federal Communications Commission (FCC) for communication equipment. These standards encompass various aspects, including performance specifications, safety requirements, and testing methodologies. The establishment of clear and consistent standards is essential for the reliable manufacturing and effective communication of photonic technologies, ensuring that products meet the necessary quality benchmarks and function seamlessly in diverse applications.

## 2.8 Skills and Workforce Development

Regulation plays a vital role in ensuring the safe and environmentally responsible use of photonic and optoelectronic technologies [30]. Safety regulations govern the use of lasers and optical radiation to protect workers and the public from potential hazards. Environmental regulations address the proper disposal and recycling of photonic components, such as LEDs and solar cells, to minimize their negative environmental impact. Manufacturers must also adhere to compliance requirements set by agencies like the Food and Drug Administration for medical devices and the Federal Communications Commission for communication equipments. The key concerns are compliance with international requirements and standards as part of photonic products' market entry and acceptance requirements, safety, performance, and environmental standards.

There are, therefore, the need to focus on a skilled workforce to enhance the photonics and optoelectronic technology systems and their uses. Core competencies in optics, electronics, and materials science are necessary to operate photonic and optoelectronic devices. Another important aspect in design of photonic structures is also the hands-on experience in the manufacturing of the corresponding components, as well as in testing of thc structures and components, both in laboratory and in real-life applications. University and technical college courses offer the basics of photonics and related areas, as well as practical experience. Also, bodies like International Society for Optics and Photonics (SPIE) and the Optical Society provide certification and annual training and education sessions to update individuals on the latest development.

## 2.9 Economic Impact and Market Trends

### *2.9.1 Market Analysis*

Knowledge of the market trends and its behavior is fundamental for the development of photonics and optoelectronics industry. Important applications are telecommunication: high-speed data transmission and network; healthcare: rapidly growing; manufacturing: laser-based manufacturing and optical sensors in smart manufacturing; consumer electronics: optoelectronics including light-emitting diodes, organic light-emitting diode displays, and camera sensors. Currently there is a constant growth in the global photonics market due to increasing demand in these sectors. Experts believe that significant growth is on the horizon as photonics is increasingly integrated into emerging technologies like 5G, self-driving cars, and smart communities.

### 2.9.2 Economic Benefits

The economic benefits are vast, spanning across production and services, with little correlation to costs. There are a number of occupations depending on the photonics industry includes research and development, production, and application engineering positions. Several advancements realized in the field of photonics help to boost the economy through increased efficiency and new revenues streams [35]. Photonics also has the advantage of reducing costs by increasing accuracy, minimizing losses during the production process, and maximizing productivity. Energy-efficient lighting and communication technologies also cut the operating expenses and add more to the sustainability. For instance, laser-based techniques in manufacturing increase accuracy and throughput, thus cutting down on scrap and rectification costs, while optical sensors allow quality checking in real-time. In healthcare, technology has been used to improve disease diagnostic techniques to enable early detection, hence sparing extra costs. In telecommunications, fiber optics supports high-speed data transmission and efficient data centers, while in automotive industry, LIDAR and optical sensors improve safety and production [36].

### 2.9.3 Investment and Funding

Investment and funding are pivotal for advancing photonics and optoelectronics technologies. The private sector heavily invests in photonics R&D to drive innovation and maintain competitive advantage [37]. Startups attract significant venture capital to develop emerging technologies like quantum photonics and integrated photonics. Government funding programs, such as those from the National Science Foundation (NSF) and the Department of Energy (DOE) in the USA, Horizon Europe in the EU, and national initiatives in Asia, support research and commercialization. Public–private partnerships, exemplified by AIM Photonics in the US and Photonics21 in Europe, and industry consortia like the Photonics Industry & Technology Development Association (PIDA) in Taiwan and the European Photonics Industry Consortium (EPIC), play a crucial role in fostering collaboration, funding, and standardization efforts.

The global optoelectronics market size was estimated at USD 6.54 billion in 2022 and it is projected to surpass around USD 21.2 billion by 2032, growing at a CAGR of 12.48% during the forecast period from 2023 to 2032. This is visualized in the form of pictorial representation in Fig. 2.6.

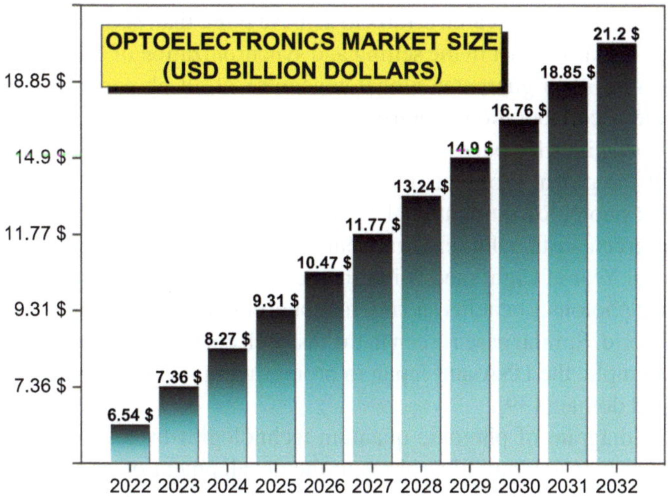


**Fig. 2.6** Estimation of optoelectronics market size

## 2.10 Collaborative Ecosystem

The collaboration between industry and academic institutions is essential for the advancement of photonics and optoelectronics [38]. These partnerships combine the complementary skills of both the sectors to foster innovation, research and commercialization. Academic organizations contribute fundamental research and new ideas in technology, while industry organizations focus on the implementation and market strategies for these technologies. There is rapid advancement of new technologies due to this synergy. Moreover, collaborations allow academic researcher to get funding, better equipment and real-life data from industries, thus increasing the potential of the research work being done. The combination of research, technical projects, and internships prepares human resources by providing students with practical experience and exposure to the real world. Some of the successful implementation of such collaborations includes the one that exists between MIT and Intel, where they collaborated in the production of the silicon photonics research in order to produce the high-speed optical interconnects for data centers. Another example is the American Institute for Manufacturing Integrated Photonics (AIM Photonics), a technology industry development initiative focused on advancing integrated photonics. Another example of such synergies at the industry level is the European Photonics Industry Consortium (EPIC) that fosters multi-partner projects, cooperation, and exchange of information between industry and academia in Europe [39].

It is also important to have a global cooperation in developing photonics and optoelectronics because it captivates the international collaboration, data exchange, and sharing and implementing the resources. International collaborations provide researchers with an opportunity to work on multi-phrenic issues because often multiple countries join research efforts to achieve more than solitary research would be able to. They also assist in the setting and synchronization of standards and legislation, in an effort to guarantee global coherence of all photonic technologies. In addition, these collaborations contribute to market development by creating new market opportunities for photonic products and technologies.

Some of the global programs include Horizon Europe, which is an EU funded program for photonics international cooperation with emphasis on health, digitization and green/sustainable environment. As a result of the proclamation of the 'International Year of Light' and 'Light-based Technologies' by the United Nations, awareness of photonics benefits and international collaboration was enhanced rapidly across the world. Similar case is seen in the research collaboration between two countries, for example, the USA and Japan in promoting innovation through partnership in the related domain [39].

The flow diagram of photonic quantum technologies from research in material systems, device design and algorithms to the development of parts, assembly, and applications in various fields of cyber-security, healthcare, artificial intelligence, and logistics is shown in Fig. 2.7 [30]. While the procurement processes of knowledge development, system specification, system development, and system integration, all contribute to the advancement of computing, communication and sensing.

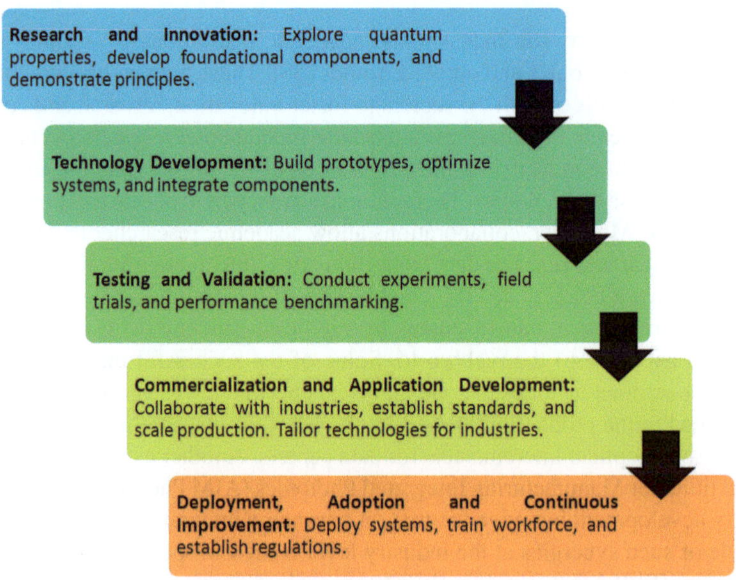


**Fig. 2.7** Development stages of photonic quantum technologies from innovation to applications across various industries

## 2.11 Challenges and Future Trends

Industry 5.0 provides a shift to the human-oriented, sustainable, and integrated industrial systems, where photonics and optoelectronics are crucial in the transformation. These technologies make it possible for innovations that increase production, efficiency and sustainability in production from different sectors. Industry 5.0 in manufacturing is defined by the integration of the advanced photonics technologies in smart factories. These innovations give manufacturers the capability for continuous monitoring, exact control, and manufacturing flexibility based on real-time market and customer requirements. Sensors driven by photonics and artificial intelligence allow for better organization of production processes, less material waste, and higher quality control, turning linear manufacturing processes into modular and flexible ones that can easily adapt to new tasks and adjust to the needs of customers [12].

The healthcare industry is undergoing a revolution through integration of photonics technologies for diagnosis, therapy and management. Molecular imaging, optical coherence tomography (OCT) and fluorescence spectroscopy ensure the clinicians get ultra-structural and functional information of tissues for early diagnosis and target treatment. Using optoelectronics in wearable applications, real-time health assessment is made possible and individuals are equipped with health information while in turn, enhancing remote patient care. The photonics technologies are revolutionizing the growth and transformation of telecommunications and data centers. Fiber optic networks by means of photonic technologies transmit high-speed data required for 5G networks, IoT devices, and smart city applications. Photonic integrated circuits improve data centers power consumption significantly, while improving data processing rates required meeting the growing demand for data-centric applications and services [17, 30, 40]. In today's automotive industry, photonics technologies are transforming the way vehicles are made safe, intelligent and connected. A number of technologies such as LIDAR for light detection and ranging and optical sensors for effective mapping and navigation for real-time self-driving cars contributing to the safer and efficient transportation systems. In the same way, photonics contributes to the implementation of smart traffic systems that help address traffic jam issues, and enhance traffic conditions in metropolitan cities. Consumer electronics are most benefited by photonics, primarily in display devices including LED and OLED. These technologies provide higher, energy-efficient luminance for display screens in mobile devices, TVs, and wearable electronics, as well as improving the user experience and expanding the opportunities of AR/VR device applications. In the future, photonics and optoelectronics are expecting much greater development. These technologies will only continue to advance with ongoing research, development, and partnership between industries and universities. International cooperation will be essential for the progress of photonics and optoelectronics to address the increasing needs for performance, sustainability, and networking in the context of Industry 5.0.

Nonetheless, despite the tremendous benefits of photonics and optoelectronics, several technical challenges must be addressed. The integration complexity is a major challenge because the incorporation of photonic and optoelectronic components in new or existing systems is costly in terms of both infrastructure and staff training. Another limitation is the cost, since such systems may be expensive to implement, especially for some industries or the initial costs of the systems may be prohibitive. Further, there are technical challenges including signal distortion, positional accuracies, and mechanical properties of materials used in photonic systems which affect the efficiency and dependability (reliability) of the photonic system.

In light of the above developments, photonics and optoelectronics in Industry 5.0 remain full of promise for the future. Quantum photonics is expected to become the next-generation communication, computation, and sensing technologies that rely on ultra-secure communication and improved quantum computing. Adoption with artificial intelligence and machine learning will take industrial systems to the next level; systems that include better and self-managed systems for improving predictive maintenance of machines, better process control, and decision-making [41]. New developments in photonic sensors and communication systems like photonic integrated circuits, as well as improved fiber optic networks, will work as enablers in the enhancement of the work precision, speed, and reliability. Further, nanophotonic will open up groundbreaking technologies such as the invisibility cloak and super-resolution microscopy, to be at the frontier of various disciplines including biomedicine and environmental science [42]. All these advancements in technology are expected to change industries and alter the development of the industry 5.0. Photonics and optoelectronics integrate with antennas to create faster and more energy-efficient communication networks for Industry 5.0 [40, 41]. They also support AI by enabling quick data processing, better decision-making, and smarter automation in industries [42, 43]. The key novelty of this work lies in its exploration of the pivotal role photonics and optoelectronics plays in shaping the technological framework of Industry 5.0. It highlights the seamless integration of these technologies with advanced automation, AI, and human-centric approaches, offering a comprehensive overview of their transformative potential in revolutionizing industries for enhanced productivity and sustainability.

## 2.12 Conclusion

Photonics and optoelectronics are fundamental to Industry 5.0 advancement with impacts on manufacturing, healthcare, telecommunications, and much more. As Industry 5.0 emphasizes the convergence of technological advancements with human-centric values, photonics and optoelectronics emerge as foundational pillars in bridging the physical and digital worlds. These technologies not only optimize industrial processes but also empower innovations in critical areas such as autonomous systems, precision medicine, and green energy solutions. These technologies improve the accuracy of operations, allow for constant supervision, and develop improved man–machine interfaces and are environmentally friendly. Smart factories that incorporate photonic and optoelectronic components are good examples of how these elements make the industrial processes more flexible and efficient. These technologies also contribute to the environmental conservation because they help in minimizing an organization's resources consumption. Laser systems, optical sensors, fibers, LEDs, and OLEDs are important technologies that facilitate a wide range of industrial applications. Laser technology offers high degree of accuracy in cutting and welding, optical sensors give accurate data on the environment, and fiber optics in data transfer that is very ultrafast. LEDs and OLEDs are beneficial to enhancing display technologies through improved and effective use of energy in the end-user point. Emerging innovations in quantum photonics, integrated photonics, and nanophotonic are paving the way for future advancements. Quantum photonics is expected to revolutionize secure communication and computational capabilities with ultra-secure networks and quantum computers. Integrated photonics promises to enhance device efficiency and miniaturization by integrating photonic components onto single chips. Nanophotonic, through developments in metamaterials and plasmonics, will enable groundbreaking applications, such as invisibility cloaks and super-resolution imaging. The economic impact of photonics and optoelectronics is substantial, contributing to job creation, economic growth, cost savings, and efficiency improvements. These technologies foster innovation, reduce operational costs, and enhance productivity across multiple sectors. As Industry 5.0 evolves, the continuous advancement of photonics and optoelectronics will drive smarter, adaptable, and sustainable industrial practices. Looking ahead, ongoing research and development will push the boundaries of photonics, leading to new materials, devices, and applications. The integration of photonics into AI-driven systems promises unprecedented levels of automation, decision-making capabilities, and resource efficiency, reinforcing the vision of sustainable and resilient industrial ecosystems. Moreover, the interdisciplinary nature of photonics and optoelectronics fosters collaborative research, enabling breakthroughs in hybrid technologies and catalyzing a new wave of disruptive innovation. The photonics market is expected to experience significant growth, driven by expanding applications in both established and emerging industries. Future trends will emphasize sustainability and human-centric design, ensuring that technological advancements deliver benefits to society and the environment.

## References

1. Marris-morini D, Vakarin V, Manel J, Liu Q, Ballabio A, Frigerio J, et al. Germanium-based integrated photonics from near- to mid-infrared applications. Nanophotonics. 2018;7(11):1781–93.
2. Das A, Rewari S, Kanaujia BK, Gupta RS. Recent technological advancement in surrounding gate MOSFET for biosensing applications—a synoptic study. SILICON. 2022;14(10):5133–43.
3. Das A, Kanaujia BK, Nath V, Rewari S, Gupta RS. Impact of reverse gate oxide stacking on gate all around tunnel FET for high frequency analog and RF applications. In: 2020 IEEE 17th India council international conference, INDICON 2020. 2020. pp. 1–6.

4. Das A, Kanaujia BK, Deswal SS, Rewari S, Gupta RS. Doping induced threshold voltage and ION/IOFF ratio modulation in surrounding gate MOSFET for analog applications. In: 2022 IEEE international conference of electron devices society Kolkata chapter (EDKCON), Kolkata, India. 2022. pp. 1–6. https://doi.org/10.1109/EDKCON56221.2022.10032940.
5. Goyal P, Srivastava G, Madan J, Pandey R, Gupta RS. Source material valuation of charge plasma based DG-TFET for RFIC applications. Semicond Sci Technol. 2022;37(9): 095023.
6. Sharma S, Goel A, Rewari S, Nath V, Gupta RS. Characterisation and analysis of Schottky-Tube FET exhibiting superior characteristic parameters. Arab J Sci Eng. 2023;48(1):907–17. https://doi.org/10.1007/s13369-022-07200-x.
7. Sharma S, Nath V, Deswal SS, Gupta RS. Analytical modelling and sensitivity analysis of Gallium Nitride-Gate Material and dielectric engineered-Schottky nano-wire fet(GaN-GME-DE-SNW-fet) based label-free biosensor. Microelectron J. 2022;129(June): 105599.
8. Das A, Rewari S, Kanaujia BK, Deswal SS, Gupta RS. Analytical characterization of a label free Si/InAs hetero-interfaced cylindrical BioFETD for biosensing applications. Micro and Nanostruct. 2025;208152. https://doi.org/10.1016/j.micrna.2025.208152.
9. Goyal P, Srivastava G, Madan J, Pandey R, Gupta RS. Analytical modelling and reliability analysis of charge plasma-assisted $Mg_2Si$/Si heterojunction doping less DGTFET for low-power switching applications. Phys Scr. 2024;99(1): 015008. https://doi.org/10.1088/1402-4896/ad1232.
10. Goyal P, Srivastava G, Madan J, Pandey R, Gupta RS. A $Mg_2Si$/Si heterojunction based dielectric modulated dopingless TFET biosensor for label free detection. Mater Sci Eng, B. 2024;304: 117356. https://doi.org/10.1016/j.mseb.2024.117356.
11. Shabbir S, Shaari A, Ul Haq B, Ahmed R, AlFaify S, Ahmed M, et al. First-principles investigations of structural parameters, electronic structures and optical spectra of 5–5- and BeO-type of $ZnO_{1-x}S_x$ alloys. Mater Sci Eng, B. 2020;262(February): 114697. https://doi.org/10.1016/j.mseb.2020.114697.
12. Grabowska S, Saniuk S, Gajdzik B. Industry 5.0: improving humanization and sustainability of Industry 4.0. Scientometrics. 2022;127(6):3117–44. https://doi.org/10.1007/s11192-022-04370-1.
13. Massaro A. Advanced electronic and optoelectronic sensors, applications, modelling and industry 5.0 perspectives. Appl Sci (Switzerland). 2023;13:4582.
14. Mourtzis D, Angelopoulos J, Panopoulos N. The future of the human–machine interface (HMI) in society 5.0. Future Internet. 2023;15(5):162.
15. Rožanec JM, Novalija I, Zajec P, Kenda K, Tavakoli Ghinani H, Suh S, et al. Human-centric artificial intelligence architecture for industry 5.0 applications. Int J Prod Res. 2022;61(20):6847–72. https://doi.org/10.1080/00207543.2022.2138611.
16. Zhang Q, Shang Q, Su R, Do TTH, Xiong Q. Halide perovskite semiconductor lasers: materials, cavity design, and low threshold. Nano Lett. 2021;21(5):1903–14.
17. Butt MA, Mateos X, Piramidowicz R. Photonics sensors: A perspective on current advancements, emerging challenges, and potential solutions (Invited). Phys Lett. 2024;516: 129633. https://doi.org/10.1016/j.physleta.2024.129633.
18. Butt MA, Kazanskiy NL, Khonina SN, Voronkov GS, Grakhova EP, Kutluyarov RV. A review on photonic sensing technologies: status and outlook. Biosensors. 2023;13(5):1–33.
19. Karunakaran C, Bhargava K, Benjamin R. Biosensors and bioelectronics. Biosens Bioelectron. Elsevier Science; 2015.
20. Huang Y, Hsiang EL, Deng MY, Wu ST. Mini-LED, Micro-LED and OLED displays: present status and future perspectives. Light: Sci Appl. 2020;9(1):105. https://doi.org/10.1038/s41377-020-0341-9.
21. Zhang W, Xu J. Advanced lightweight materials for automobiles: a review. Mater Des. 2022;221: 110994.
22. Mabkhot MM, Al-Ahmari AM, Salah B, Alkhalefah H. Requirements of the smart factory system: a survey and perspective. Machines. 2018;6(2):1–22.
23. Ma X, Tao F, Zhang M, Wang T, Zuo Y. Digital twin enhanced human-machine interaction in product lifecycle. In: Procedia CIRP. Elsevier B.V.; 2019. pp. 789–93. https://doi.org/10.1016/j.procir.2019.04.330.

24. Rosen MA, Kishawy HA. Sustainable manufacturing and design: concepts, practices and needs. Sustainability. 2012;4(2):154–74.
25. Das A, Rewari S, Kanaujia BK, Deswal SS, Gupta RS. Numerical modeling of a dielectric modulated surrounding-triple-gate germanium-source MOSFET (DM-STGGS-MOSFET)-based biosensor. J Comput Electron. 2023;22(2):742–59. https://doi.org/10.1007/s10825-023-02008-w.
26. Yadav S, Das A, Rewari S. Dielectrically-modulated GANFET biosensor for label-free detection of DNA and Avian influenza virus: proposal and modeling. ECS J Solid State Sci Technol. 2024;13(4): 047001. https://doi.org/10.1149/2162-8777/ad3364.
27. Das A, Rewari S, Kanaujia BK, Deswal SS, Gupta RS. Ge/Si interfaced label free nanowire BIOFET for biomolecules detection-analytical analysis. Microelectron J. 2023;138:105832. https://doi.org/10.1016/j.mejo.2023.105832.
28. Das A, Rewari S, Kanaujia BK, Deswal SS, Gupta RS. Analytical investigation of a triple surrounding gate germanium source metal–oxide–semiconductor field-effect transistor with step graded channel for biosensing applications. Int J Numer Model Electron Networks Devices Fields. 2023;36(6): e3106. https://doi.org/10.1002/jnm.3106.
29. Das A, Rewari S, Kanaujia BK, Deswal SS, Gupta RS. Physics based numerical model of a nanoscale dielectric modulated step graded germanium source biotube FET sensor: modelling and simulation. Phys Scr. 2023;98(11): 115013. https://doi.org/10.1088/1402-4896/acf4c9.
30. Slussarenko S, Pryde GJ. Photonic quantum information processing: a concise review. Appl Phys Rev. 2019;83: 016001.
31. Cheben P, Halir R, Schmid JH, Atwater HA, Smith DR. Subwavelength integrated photonics. Nature. 2018;560:565–72. https://doi.org/10.1038/s41586-018-0421-7.
32. Fulmore JS, Geiger BF, Werner KA, Talbott LL, Jones DC. Nanophotonics: interactions, materials, and applications. J Phys Chem B. 2009;104(32):7577–87.
33. Das A, Rewari S, Kanaujia BK, Deswal SS, Gupta RS. Analytical modeling and doping optimization for enhanced analog performance in a Ge/Si interfaced nanowire MOSFET. Phys Scr. 2023;98(7):74005. https://doi.org/10.1088/1402-4896/acde16.
34. Pirovano A, Lacaita AL, Spinelli AS. Two-dimensional quantum effects in nanoscale MOSFETs. IEEE Trans Electron Devices. 2002;49(1):25–31.
35. Shaker LM, Al-Amiery A, Isahak WNRW. Optoelectronics' quantum leap: unveiling the breakthroughs driving high-performance devices. Green Technol Sustain. 2024;2(3): 100111. https://doi.org/10.1016/j.grets.2024.100111.
36. Li N, Ho CP, Xue J, Lim LW, Chen G, Fu YH, et al. A progress review on solid-state LiDAR and nanophotonics-based LiDAR sensors. Laser Photonics Rev. 2022;16(11):1–24.
37. Liu CC, Cheng AC, Chen SH. A study for sustainable development in optoelectronics industry using multiple criteria decision making methods. Technol Econ Dev Econ. 2017;23(2):221–42.
38. Ponraj JS, Xu ZQ, Dhanabalan SC, Mu H, Wang Y, Yuan J, et al. Photonics and optoelectronics of two-dimensional materials beyond graphene. Nanotechnology. 2016;27: 462001.
39. Yang Y, Badloe T, Song D, Park S, Rho J. Building an optics and photonics research ecosystem in South Korea: Collaborative innovation between academia and industry. Light: Sci Appl. 2023;12:1–4.
40. Sharma S, Rewari S, Nath V, Deswal SS, Gupta RS. Schottky barrier double surrounding gate MOSFET for high frequency implementation. In: 2020 5th IEEE International conference on recent advances and innovations in engineering (ICRAIE). 2020. pp. 5–8.
41. Wu Y, Liu X, Pandey A, Zhou P, Dong WJ, Wang P, et al. III-nitride nanostructures: Emerging applications for Micro-LEDs, ultraviolet photonics, quantum optoelectronics, and artificial photosynthesis. Prog Quantum Electron. 2022;85: 100401. https://doi.org/10.1016/j.pquantelec.
42. El Srouji L, Krishnan A, Ravichandran R, Lee Y, On M, Xiao X, et al. Photonic and optoelectronic neuromorphic computing. APL Photonics. 2022;7: 051101.

# Chapter 3
# Fundamentals of Photonics and Optoelectronics in Modern Industrial Processes

**Amit Das, Preeti Goyal, Swati Sharma, Anjana Bhardwaj, Kaushik Das, and R. S. Gupta**

**Abstract** This chapter explores the fundamental principles, industrial applications, and future trends of photonics and optoelectronics. It highlights their role in modern industries such as manufacturing, telecommunications, healthcare, and automation, emphasizing their impact on high-speed data transmission, precision engineering, and advanced sensing technologies. Key developments that have been comprehensively discussed include photonic integrated circuits, silicon photonics, and laser-based manufacturing systems, which drive efficiency and innovation in Industry 4.0. The chapter also examines emerging trends like AI-driven photonics, quantum photonics, and sustainable low-power photonic devices. Despite their transformative potential, challenges such as cost, workforce training, and standardization must be addressed to enable widespread adoption and integration into future industrial advancements.

**Keywords** Absorption · Industry 4.0 · LASER · Optoelectronics · Photonics

All the authors have equally contributed to this work.

A. Das (✉) · A. Bhardwaj (✉)
Department of Electronics and Communication Engineering, GLA University, Mathura, Uttar Pradesh, India
e-mail: amitofficial7492@gmail.com

A. Bhardwaj
e-mail: anjana0889@gmail.com

P. Goyal (✉) · S. Sharma (✉) · R. S. Gupta
Department of Electronics and Communication Engineering, Maharaja Agrasen Institute of Technology, New Delhi, India
e-mail: preetigoyal.s@gmail.com

S. Sharma
e-mail: swati1986sharma@gmail.com

K. Das (✉)
Department of Electronics and Communication Engineering, National Institute of Technology Jamshedpur, Adityapur, Jharkhand, India
e-mail: daskk62@gmail.com

 
V. Bhardwaj et al. (eds.), *Photonics and Optoelectronics in Industry 5.0*, Advances in Optics and Optoelectronics, https://doi.org/10.1007/978-981-96-7594-4_3

## 3.1 Introduction

Photonics with optoelectronics stands as a fundamental modern industrial technology that has transformed various industrial operations, particularly in telecommunications, healthcare, aerospace, manufacturing, and automation sectors. It has enabled innovations in high-speed data transmission, precision engineering, and advanced sensing capabilities. The digital transformation in industries finds its foundation through photonics technology which empowers smart factories in addition to Industry 4.0 implementation [1, 2]. Real-time production line surveillance through optical sensors and imaging systems achieves improved operational performance with fewer errors. It is now widely used in various precise cutting operations along with advanced manufacturing processes because it delivers exceptional control and accuracy. Photonic integrated circuits (PICs) that incorporate optical and electronic elements in their components miniaturistically increase data speed and advanced electrical computation operations' transfer and processing. The expansion of silicon photonics is becoming more and more pronounced because it provides the ability to integrate with the existing semiconductor technologies, which makes quantum computing and energy-efficient communication networking possible. The ongoing development in photonics and optoelectronics is determined by scientific achievements and industrial business needs. The intellectual convergence enables the use of bioopto sensors, high-power lasers, quantum photonics, and other advanced technologies in the photonics' industrial branches outside the traditional optics industry. With deep scientific and technological scrutiny, this chapter considers astronomic micro-level details and demonstrates the fundamental principles of photonics and optoelectronics and their industrial implementations. Their assessment will be done retrospectively, examining the information of inventing and creating new materials for these technological masterpieces and information about published articles regarding the prospects and main achievements in the provided fields.

Through the regulation of light (photonics and optoelectronics) and its subsequent transformation into various technological uses, contemporary industrial methods have been changed through light-based technologies and electronics. This synchronization of both sectors caused critical progress in automation, telecommunications, healthcare, aerospace, and manufacturing. In the last few years, the combination of photonics with Industry 4.0 has enabled the development of smart factories to enhance precision, efficiency, and automation by using optical sensors, fiber optics, and laser-based manufacturing systems. In addition, the emergence of photonic integrated circuits and silicon photonics promise revolutionary advances in high-speed data processing and communication. This chapter focuses on photonics and optoelectronics, including their elementary physics, primary devices, industrial applications, and other aspects that pertain to their latest developments. The chapter discusses developments, achievements, illustrative examples, and new research that trace the contribution of these two fields (sciences) to change industrial systems in the modern world.

## 3.2 Fundamental Physics and Working Principles

### *3.2.1 Light-Matter Interaction*

The light and matter interaction is the foundation of photonics and optoelectronics. This interaction takes place in multiple fundamental processes like absorption, emission, reflection, refraction, scattering, and photoelectric effect. Lasers, photodetectors, and even advanced optical communication systems works in accordance to these principles.

***Absorption and Emission***: Absorption occurs when materials capture incident photons, leading to higher excitation of electrons to more elevated energy states [3]. This principle is fundamental in photodetectors and solar cells. Emission describes the energy release that matter undergoes wherein it is expelled in photons at high temperatures. Spontaneous emission accompanies the light-emitting diodes, whereas their laser counterpart comes from stimulated emission.

***Reflection and Refraction***: Reflection is associated with the bouncing of light and the formation of an image when it hits a medium and comes across the surface of a material. Light bending by refraction as it cuts through different media is another phenomenon. These phenomena play essential roles in the design of lenses, optical fibers, and commercial coatings that control the loss of light for industrial, biomedical, and other applications.

***Scattering and Raman Effect***: Scattering occurs when light appears visibly agitated due to irregularities in a medium, causing deflections that impact aseismic imagery or imaging-based sensing. One form of scattering, known as Raman scattering, is of great significance in analyzing materials in spectroscopy [4].

***Photoelectric Effect***: When light shines on a material with sufficient energy, it releases electrons, which is called the photoelectric effect. This concept enables the use of photovoltaic cells in solar energy and photodetectors in imaging and sensing devices [5].

### *3.2.2 Semiconductor Physics*

Optoelectronic devices rely on semiconductors to absorb, emit light, and convert it into an electrical signal. Its electronic band structure defines the optical characteristics of semiconductors, which is why compounds such as silicon, gallium arsenide, and indium phosphide are used in LEDs, lasers, and photodetectors [6].

***Band Structure and Energy Gaps***: Like insulators, semiconductors have a valence band where electrons are bound and a conduction band in which they can move freely. A gap separates these bands. The magnitude of this bandgap of semiconductors defines the range of light wavelengths that can be absorbed or emitted, setting boundaries for infrared, visible, and ultraviolet technology.

***Carrier Generation and Recombination***: In semiconductors, photons are absorbed, and electron–hole pairs are formed. These pairs are beneficial in photodiodes and solar cells. In LED and laser diodes, electro-luminescence and stipulated emission form the device operation basis.

***Doping and PN Junctions***: Doping introduces impurities into semiconductors to modify their electrical characteristics [7–10]. The formation of PN junctions, where p-type and n-type materials meet, is fundamental to the function of LEDs, laser diodes, and photodetectors, allowing for controlled electron movement and light emission.

***Quantum Wells and Nanostructures***: Advancements in nanotechnology have led to the development of quantum wells, quantum dots, and nanostructures that enhance optoelectronic performance. Quantum wells confine electrons within ultra-thin layers, improving laser efficiency, while quantum dots allow for tunable emission properties, revolutionizing display technologies and biomedical imaging.

## 3.3 Fundamentals of Photonics

Photonics is the science and technology of generating, controlling, and detecting light. It forms the foundation of numerous modern applications, including fiber-optic communication, laser technology, and imaging systems. The study of photonics integrates elements of optics, quantum mechanics, and material science, facilitating the advancement of highly efficient photonic devices and circuits.

Photonics exploits the dual nature of light—its ability to behave both as a wave and a particle. This characteristic is essential in designing optical components such as lenses, mirrors, diffraction gratings, and semiconductor-based optoelectronic devices like LEDs, lasers, and photodetectors. Modern photonics also encompasses nonlinear optical effects, where high-intensity light can alter the optical properties of a medium, leading to applications in optical signal processing, high-speed communication, and laser machining. Recent advancements in photonic integrated circuits (PICs) have led to miniaturized and highly efficient optical components that significantly improve data transmission speeds and reduce energy consumption. These circuits integrate multiple optical functions on a single semiconductor chip, enabling new possibilities for computing, artificial intelligence, and medical diagnostics.

Materials used in photonics can be broadly categorized based on their optical properties and applications. Dielectric materials, such as fused silica and optical polymers, exhibit low absorption and are commonly used in lenses, optical fibers, and waveguides. Metals like silver and aluminum have reflective materials essential for mirrors, coatings, and plasmonic applications where light manipulation at the nanoscale is required. Semiconductor materials such as silicon (Si), gallium arsenide (GaAs), and indium phosphide (InP) are integral to optoelectronic devices, including LEDs, laser diodes, and photodetectors [11–13]. Nonlinear optical materials exhibit optical properties that change with light intensity and are utilized in frequency conversion, optical switching, and laser modulation. The development

**FUNDAMENTALS OF PHOTONICS**

**1 CORE PRINCIPLES**

- Wave Particle Duality – Light behaves as both a wave and a particle (photon).
- Light-Matter Interaction – Processes like absorption, emission, and scattering.
- Optical Phenomena – Key effects that define how light behaves.

**2 OPTICAL PHENOMENA**

- Interference & Diffraction – Wave interactions that affect light patterns.
- Reflection & Refraction – Light bouncing off and bending through surfaces.
- Polarization & Coherence – Controlling light orientation and phase.

**3 KEY TECHNOLOGIES**

- Lasers & LEDs – Devices for light generation and applications.
- Optical Fibers – High-speed data transmission using light.
- Photodetectors – Sensors that detect and measure light.

**4 INDUSTRIAL APPLICATIONS**

- Telecommunications – Fiber optics for internet and communication.
- Medical Imaging – Optical techniques for non-invasive diagnostics.
- Manufacturing & Automation – Laser cutting, optical sensors in Industry 4.0.

**Fig. 3.1** Synoptic overview of fundamentals of photonics

of metamaterials—the materials designed with tailored optical properties—allowed previously unachievable control over light propagation and created such things as invisibility cloaks, superlenses, and ultrafast photonic processors. Figure 3.1 shows the synoptic overview of the fundamentals of photonics.

The optical phenomena underpin various industrial applications, enabling advancements in imaging, sensing, manufacturing, and communication technology. Key optical phenomena utilized in industrial processes include:

- **Interference**: Interference combines two or more light waves that meet, leading to places of constructive and destructive interference as defined by: This phenomenon finds immense applications in measurements, holography, and film coatings. Interferometry finds application in various areas where, for instance, material volume has to be measured, surface profile has to be determined, and mechanical structures should be analyzed for strain gauge performance. Among industrial applications and spectroscopy, Michelson and Fabry-Pérot interferometers define the most essential tools.
- **Diffraction**: Diffraction refers to the bending and spreading of light waves due to an obstacle or through an aperture. This phenomenon is essential in laser optical filtering, engraving, and spectrometer usage. The diffraction grating uses these devices with their periodic features to cause the separation of wavelengths for accurate measurement in computing, telecommunications, and fiber optics. Newer diffraction techniques can provide focused images with high resolution in biomedical optics and semiconductor lithography.
- **Polarization**: Polarization describes the orientation of light wave oscillations. This property is widely used in liquid crystal displays (LCDs), optical communication, and stress analysis in materials. Polarimetry utilizes polarized light to

examine the chemical makeup of substances and, in industrial inspection systems, to identify irregularities and defects on the surfaces of transparent materials.

- **Coherence**: Coherence is the term used to describe the relationship in the phase of light waves. Highly coherent lasers are used in precision engineering, metrology, and medical imaging. More elaborate coherence techniques, like optical coherence tomography (OCT), are used for high-resolution cross-sectional imaging for medical diagnostics and inspection of materials in industry.
- **Total Internal Reflection (TIR)**: This happens when light is incident at some angle to the normal, and some part of the light is reflected totally instead of being refracted. This is the basic principle of optical fibers, which is where light is transmitted with little loss for very long distances. This technology is significant for telecommunications, remote sensing, and endoscopy.
- **Raman Scattering**: Raman scattering is where inelastic photons scattered from a system undergo specific energy shifts proportional to molecular vibrations to provide a molecular fingerprint of the substance. Raman scattering is widely used for chemical analysis, pharmaceutical industry, and industrial process control for contamination detection, composite materials characterization, and product quality control.

## 3.4 Core Principles of Optoelectronics

Optoelectronics is a technology discipline related to the interaction between light and electronic systems. The devices used for generating, manipulating, and detecting optical signals fall under this category. These devices are fundamental to modern communication, sensing, and imaging technologies. Figure 3.2 shows the synoptic overview of the fundamentals of photonics.

Optoelectronic devices work based on electron-photon interactions within semiconductors. Such interactions enable various functions, including light emission, photodetection, and signal modulation. During the last few years, major nanophotonics and quantum optics efforts have been motivated by the increasing demand for high-speed, low-power optoelectronic devices to create more efficient miniaturized devices. Key optoelectronic devices include:-

- **Lasers**: Lasers lie at the heart of optoelectronics, generating coherent, monochromatic light to be used in material processing, medical diagnostics, and fiber-optic communications. The creation of quantum cascade and fiber lasers has opened up a broader spectrum of wavelengths available for both industrial and scientific applications.
- **Light Emitting Diodes (LEDs)**: LEDs are highly efficient and very flexible light sources for use in displays, automotive lighting, and energy-efficient illumination. There have been sensational advances in the field of organic LEDs (OLEDs) and micro-LEDs that make the performance and the scope of applications available better further.

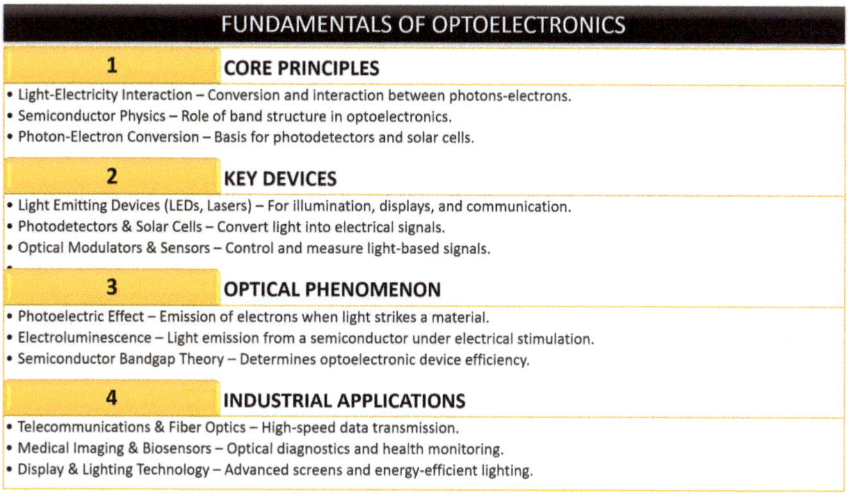


**Fig. 3.2** Synoptic overview of fundamentals of optoelectronics

- **Photodetectors**: Devices for transforming light into electrical signals act accordingly to be used in cameras, biomedical imaging, and environmental monitoring. Recent advancements concerning avalanche photodiodes (APDs) and single-photon detectors prove how sensitivity and detection accuracy have improved significantly.
- **Optical Modulators**: Modulators control the properties of light signals, including amplitude, phase, and polarization, making them essential for high-speed data transmission in optical networks.
- **Solar Cells**: Photovoltaic apparatuses transform solar energy into electricity and are essential in renewable energy sources. Perovskite and multi-junction solar cells have achieved spectacular enhancements in efficiency values for years.

The ongoing miniaturization and integration of optoelectronic devices have led to the rise of photonic computing, where light-based logic and memory components promise ultrafast data processing with lower power consumption. As research continues, the fusion of optoelectronics with artificial intelligence and quantum technologies will drive the next generation of innovations in industrial automation, healthcare, and communication. Materials in photonics include: Dielectrics (used in lenses and optical coatings); metals (employed in mirrors and plasmonic devices); semiconductors (essential for LEDs, lasers, and photodetectors); and photonic crystals (used in optical computing and sensing).

## 3.5 Industrial Applications of Photonics and Optoelectronics

Photonics and optoelectronics are essential in numerous industrial applications, driving efficiency, automation, and innovation across various sectors [14]. The ability to manipulate light with high precision has led to significant advancements in manufacturing, telecommunications, medical imaging, security, and environmental monitoring. As industries move toward digital transformation and Industry 4.0, the adoption of photonic and optoelectronic technologies continues to expand, reshaping traditional production methods and enabling real-time monitoring and control systems. Figure 3.3 shows the key application areas of photonics and optoelectronics.

### 3.5.1 *Manufacturing and Material Processing*

The integration of photonic technologies in manufacturing and material processing has revolutionized industrial production. Laser-based processing, including cutting, welding, engraving, and surface treatment, offers unparalleled precision, speed, and minimal material waste. Industrial photonics technologies are crucial in manufacturing, delivering precision and efficiency across various applications. Laser cutting and welding enable highly controlled machining of metals, plastics, and ceramics, which are widely used in the automotive, aerospace, and consumer electronics industries. Additive manufacturing and 3D printing utilize photonics-based layer-by-layer deposition and UV curing for complex, high-accuracy structures. Optical metrology techniques like interferometry and laser scanning ensure real-time, non-contact quality control, reducing defects and waste. Holographic imaging enhances security applications, industrial design, and non-destructive testing with superior imaging

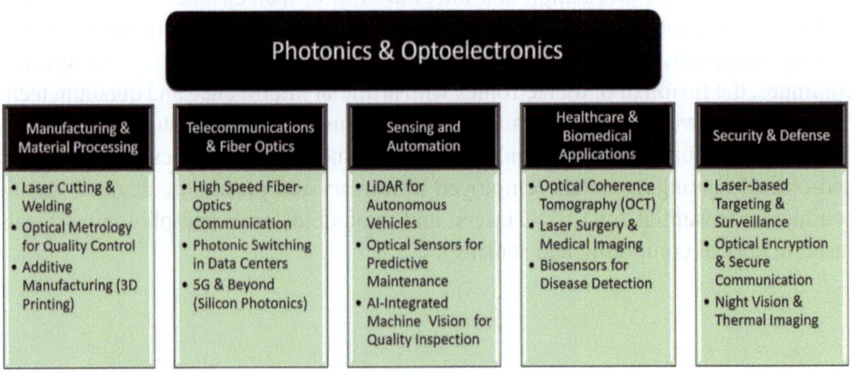


**Fig. 3.3** Key application areas of photonics and optoelectronics

capabilities. These advancements collectively improve manufacturing precision, reliability, and efficiency.

### 3.5.2 Telecommunications and Fiber Optics

Optical fiber technology has become the backbone of modern telecommunications, enabling high-speed data transmission over vast distances with minimal signal loss. Photonics and optoelectronics continue to push the boundaries of data communication, ensuring faster and more efficient connectivity solutions. Photonics is the backbone of modern communication, enabling high-speed data transfer and network efficiency. Fiber-optic communication transmits vast amounts of data at the speed of light, with dense wavelength division multiplexing (DWDM) enhancing capacity by allowing multiple signals to travel simultaneously. Photonic switches improve data routing in fiber-optic networks, optimizing data center operations while reducing power consumption. The advancement of 5G and beyond relies on photonic technologies, with silicon photonics enabling compact, high-bandwidth transceivers for low-latency wireless communication. Additionally, optical interconnects replace traditional copper wiring in high-performance computing, reducing heat dissipation and boosting processing speeds for AI and cloud computing.

### 3.5.3 Sensing and Automation

Photonics-based sensing technologies have dramatically impacted industrial automation by enhancing efficiency, safety, real-time monitoring, etc. Optic sensors can make accurate measurements—temperature, strain, pressure, and chemical composition—across the board in every industry. LiDAR technology enables precise distance measurement and environmental mapping for autonomous vehicles, robotics, and geospatial applications, playing a critical role in self-driving cars and drones [15]. High-sensitivity optical sensors monitor essential industrial parameters such as temperature, structural integrity, and chemical composition, aiding in predictive maintenance to reduce downtime and boost efficiency. Additionally, machine vision systems, which integrate optical imaging and AI, enhance quality control by detecting defects in real-time across industries like electronics assembly, pharmaceuticals, and semiconductor manufacturing. In the biomedical field, optical techniques such as optical coherence tomography (OCT) and fluorescence imaging offer non-invasive, high-resolution visualization of biological tissues, significantly improving medical diagnostics and treatment planning.

### 3.5.4 Smart Manufacturing and Industry 4.0

The combination of photonics technology with optoelectronics and automated processes is transforming modern manufacturing through the creation of innovative factory systems that improve efficiency and precision while boosting system reliability. Advanced optical sensing technologies provide essential support for real-time monitoring by collecting precise data to maintain predictive maintenance and reduce unexpected downtime. Implementing high-precision sensors enhances quality control through defect detection and production process optimization, leading to increased yields and less material waste. Optical and photonic systems drive automated manufacturing processes that decrease human intervention and boost productivity and process consistency [2].

Beyond process optimization, smart manufacturing leverages sensor-driven insights and predictive analytics for data-driven decision-making. These technologies enable adaptive and flexible production systems, allowing manufacturers to swiftly respond to changing market demands, support mass customization, and facilitate rapid prototyping. By integrating Industry 4.0 technologies, factories achieve greater operational efficiency, improved product quality, and a more sustainable approach to manufacturing, ensuring long-term competitiveness in the evolving industrial landscape.

## 3.6 Practical Examples and Applications

***Optical Communication***—Photonics have made what can be considered one of the most radical applications in fiber-optic communications, which has revolutionized how the world connects. Conventional copper-based networks, decidedly constraining in terms of bandwidth and the distance over which electric signals can be transmitted, meant that data speeds ended up being slower, not to mention high signal loss [16]. DWDM is the method that optimizes data transmission capacity into fiber-optic systems. Several companies perform a function of the deployment of gigabit-speed Internet through the usage of optic technology, including Google Fiber, AT&T, and Huawei. With terabits of data that can now be sent streaming through a solitary optical fiber, cloud computing, video streaming, and 5G networks, among others, will have the appropriate speed to use them to the slightest degree possible. All this will be made available perfectly healthy for the rightful furtherance of low-latency networking, permitting AI-driven applications, self-driven automobiles, and the subsequent 6G evolution.

***Laser Manufacturing***—Automotive industry in the automotive sector, laser technology has grown into a household name in precision manufacturing. Tesla's Gigafactory is already famous for its automation, from cutting and welding with lasers in the manufacture of electric vehicles. This is unlike conventional mechanical cutting; laser processing delivers very high accuracy with minimum material wastage and results in much quicker production. High-power lasers also weld battery cells in the EV battery assembly process with sensitivity to the order of microns that error out for longevities and safety. BMW and Toyota use ultrafast, femtosecond lasers to cut composite lightweight materials to make the cars more fuel-efficient at the same time as maintaining structural integrity. All this transition has been made possible by laser-based manufacturing in increasing production rates, lowering operating costs, and enhancing product reliability under Industry 4.0 [17].

***Biomedical Imaging***—Optical coherence tomography (OCT) in healthcare medical imaging has progressed remarkably with the emergence of optical coherence tomography, an optoelectronic device capable of high-resolution cross-sectional imaging of biological tissues. Its most typical use is in the field of ophthalmology, where it is used to diagnose glaucoma, macular degeneration, and diabetic retinopathy. Real-time, non-invasive imaging with micro-precision level capability is done in hospitals and clinics to identify structural abnormalities of the eye. Because of its ability to see tissue layers below the surface, OCT is highly beneficial to the early intervention of diseases, significantly improving patient health. Apart from ophthalmology, OCT is also employed in cardiology for imaging arterial plaques, which helps to prevent heart diseases. With the addition of artificial intelligence to assist in the IN OAG, the overall diagnostic accuracy is enhanced, and the false positive rate is decreased significantly, enabling targeted treatment [18–25]. It can also be integrated with transistor-based biosensors.

***AI-Integrated Photonic Sensors in Smart Manufacturing***—The emergence of smart factories led to the accelerated adoption of AI-based optical sensors for real-time quality control and predictive maintenance. Advanced manufacturers like Siemens, Intel, and BMW implement machine vision systems with high-resolution photonics sensors that inspect semiconductor wafers, micro-defects in electronic circuits, and material properties in real-time. These optical sensors can function with nanometer accuracy, allowing for unparalleled quality monitoring in high-volume production settings [5]. For example, Intel's chip fabrication plants use infrared photonic sensors to find atomic imperfections within silicon wafers, lower chip failure rates, and increase yield. Moreover, LiDAR-based optical sensors within industrial automation improve robotic assembly line efficiency through object recognition, distance measuring, and environmental mapping. Combining AI and photonics resulted in superior manufacturing outcomes, lower operational downtime, and greater productivity, thus enabling modern industry [15]. Table 3.1 briefly discusses (synoptic review) about the related work in photonics and optoelectronics in the corresponding literature.

**Table 3.1** Brief survey of the work in the corresponding literature

| | | |
|---|---|---|
| Yariv et al. [26] **Year**: 2007 | Photonics: Optical electronics in modern communications | This work provides a comprehensive overview of photonics, exploring fundamental principles, optoelectronics, and communication systems. It also covers advanced topics like Stokes parameters, wave propagation matrix methods, and fibers' nonlinear optical effects |
| Andrea et al. [16] **Year**: 2019 | Additive manufacturing: applications and directions in photonics and optoelectronics | This work explores the transformative impact of additive manufacturing on photonics and optoelectronics, enabling the creation of complex 3D architectures. It highlights recent advancements in 3D printing for fabricating LEDs, lasers, waveguides, and sensors. As the field continues to evolve, research focuses on printable optical materials, multi-material processing, and high-resolution, large-volume printing, paving the way for groundbreaking innovations |
| Sharma et al. [27] **Year**: 2024 | Photonics for AI and AI for photonics integration: Materials and characteristics | This work examines the convergence of photonics and AI, exploring the integration of photonics in AI systems and AI-driven advancements in photonics technology. It discusses key photonic materials for AI applications and how machine learning and neural networks optimize photonic systems. Additionally, it includes a sensitivity analysis and refractive index calculation of twin-core photonic crystal fiber (PCF). The synergy between photonics and AI holds great potential for advancing information and optimization sciences |
| Poonguzhali et al. [28] **Year**: 2022 | Design and performance analysis of smart photonic sensors for industrial applications | This paper introduces a 2D-PhC resonator-based nanosensor for toxic gas detection, leveraging hexagonal and square crystal lattices. The sensor achieves high sensitivity and low detection limits by detecting refractive index variations. Simulations demonstrate strong performance with a high-quality factor (QF) and enhanced sensitivity, making it suitable for industrial safety, healthcare, and environmental monitoring applications |

## 3.7 Case Study

***Case Study A: Photonics in Smart Cities (Singapore's Intelligent Traffic System Background)***

**Background**: As urban populations grow, traffic congestion has become a critical challenge for metropolitan cities. Singapore, known for its technological advancements, has implemented an AI-driven intelligent traffic management system using LiDAR sensors, optical cameras, and fiber-optic networks to optimize urban mobility. The goal is to reduce congestion, enhance road safety, and minimize carbon emissions by enabling real-time traffic monitoring and dynamic signal adjustments. Implementation: To tackle urban congestion, Singapore's Land Transport Authority (LTA) deployed a network of LiDAR-based traffic sensors across the city. These sensors continuously scan road conditions, vehicle density, and pedestrian movement with millimeter precision, enabling accurate data collection. AI-powered optical cameras analyze pedestrian crossings, detect traffic violations, and identify accident-prone zones. The collected data is transmitted via high-speed fiber-optic communication networks to traffic control centers, where AI algorithms process the information and dynamically adjust traffic signals in real-time. Moreover, innovative road lighting systems utilizing photonic sensors automatically adjust brightness based on traffic flow and weather conditions, reducing unnecessary energy consumption. Autonomous vehicles also benefit from real-time vehicle-to-infrastructure (V2I) communication, improving navigation and reducing collision risks [29].

**Results and Impact**: Implementing photonics-based innovative traffic management has led to several measurable improvements.

- 30% reduction in peak-hour congestion, improving travel time efficiency.
- Lower carbon emissions, as vehicles spend less time idling at traffic signals.
- Enhanced road safety, with a 15% decrease in pedestrian-related accidents.
- Energy-efficient urban lighting, reducing electricity consumption by 20%.

This case study highlights the transformative role of photonics in urban infrastructure, demonstrating how optical technologies can create sustainable, AI-driven smart cities. As more cities adopt similar models, the integration of photonic sensing, real-time AI analytics, and fiber-optic networking will pave the way for next-generation intelligent transportation systems.

***Case Study B: Photonics in Aerospace—NASA's Optical Communication System***

**Background**: Traditional radio-frequency (RF) communication has long been the primary method for transmitting data between space missions and Earth. However, as space exploration expands, RF communication struggles with limited bandwidth, higher latency, and increasing power consumption. To address these challenges, NASA has developed and deployed Laser Communication Relay Demonstration (LCRD), a photonics-based optical communication system that significantly enhances data transmission rates for deep-space missions [30].

**Implementation**: NASA's LCRD utilizes laser-based free-space optical communication to transmit high-bandwidth data between satellites and ground stations. Laser beams are highly directional, unlike RF signals, which spread over larger areas, ensuring higher data rates with minimal signal loss. The system comprises:-

- Space-based laser transmitters: These high-powered optical lasers transmit data through space 100 times faster than traditional RF systems.
- Optical ground stations: Equipped with adaptive optics and large-aperture telescopes, these stations receive high-fidelity laser signals and convert them into usable data.

- Adaptive beam steering: The system continuously adjusts laser alignment to compensate for atmospheric distortions and spacecraft movement.

One of the most notable missions using this technology is the Lunar Laser Communication Demonstration (LLCD), which successfully transmitted data from the Moon to Earth at 622 megabits per second (Mbps)—a groundbreaking speed compared to RF transmission.

**Results and Impact**: Implementing photonics in aerospace has led to several measurable improvements.

- 100× faster data transmission: Compared to conventional RF communication, enabling high-resolution imaging and real-time video streaming from space.
- Lower power consumption: Optical communication requires less energy, which is crucial for deep-space missions.
- Reduced signal interference: Unlike RF signals, which suffer from spectrum congestion, laser communication provides a dedicated, interference-free channel.
- Enhanced deep-space exploration: Future Mars and interplanetary missions will rely on photonic communication to transmit terabytes of scientific data back to Earth.

**Future Outlook**: NASA and private aerospace companies like SpaceX and Blue Origin are actively developing next-generation optical communication networks for future Moon and Mars colonization efforts. As quantum photonics and AI-integrated optical systems advance, laser communication could become the backbone of interplanetary internet connectivity, revolutionizing space exploration [1, 31].

## 3.8 Future Trends and Challenges

Emerging trends are shaping the future of photonics and optoelectronics through enhancements in performance and efficiency with broader industry applications. Integrated photonics enable photonic circuits to shrink, which results in optical systems that are compact and efficient and have high-speed performance. Machine learning applications within AI-driven photonics are transforming imaging techniques and defect detection, plus process automation through enhanced optical data analysis accuracy. Quantum photonics stands to revolutionize fields by enabling ultra-secure communication systems, advanced quantum computing, and highly sensitive sensing technologies. Sustainable practices benefit from low-power photonic devices, which enhance industrial energy efficiency and reduce environmental effects through greener manufacturing methods. The broad adoption of photonics and optoelectronics requires addressing multiple existing challenges. Reducing production and integration expenses is essential because these high costs restrict accessibility and scalability. Engineers and technicians need training to master specialized skills for developing and maintaining advanced photonic systems. Standardization must be established to achieve interoperability among photonic technologies and enable smooth integration with existing industrial and communication infrastructures. The future success of photonics and optoelectronics depends on solving these critical obstacles.

## 3.9 Conclusion

Photonics and optoelectronics have transformed industries, from manufacturing and telecommunications to sensing and automation. They play a key role in modern industrial production, enabling manufacturing, communication, and automation breakthroughs. While the human brain keeps advancing, those technologies, along with artificial intelligence, quantum computing, and practices aimed at sustainability, gradually give rise to more innovations that will make the future of industry and technology. Their ability to precisely control light has enabled new efficiency, accuracy, and sustainability levels in various applications. As the research progresses, new developments like quantum photonics, silicon photonics, and AI-integrated optical systems will likely gain more momentum in the coming years. The development of ultrafast optical computers, energy-efficient communication systems, and high-resolution biomedical imaging shows the relentless advancement of the field and its impact on society. The future of photonics is bright, even with the challenges in cost-effectiveness, restricted materials, and issues in integration with existing technologies. Any new developments in smart manufacturing and intelligent automation will require interdisciplinary collaboration and further investment in photonics R&D. The full potential of advanced communication technologies relies on the novel approaches offered through photonics and optoelectronics and, in turn, serves as the foundation for more innovation and productivity in various industries in the future. In Industry 4.0, their contribution will grow more significantly to enhance efficiency, precision, and productivity.

Technological advances in photonics and optoelectronics have recently boosted the performance of manufacturing, telecommunications, sensing, and computing industries. Further exploration of these fields may yield unprecedented developments, including the design of integrated photonic chips, AI-powered optical systems, and a new branch of science—quantum photonics. However, cost, scalability, and standardization are a few concerns that must be addressed before realizing their full potential. New technological and sustainable advancements will be made possible by focusing on improving efficiency, and the next stage of the industrial revolution will be achievable through the evolution of photonics and optoelectronics.

## References

1. Slussarenko S, Pryde GJ. Photonic quantum information processing: a concise review. Appl Phys Rev. 2019;83: 016001.

2. Massaro A. Advanced electronic and optoelectronic sensors, applications, modelling and industry 5.0 perspectives. Appl Sci (Switzerland). 2023;13:4582.
3. Liu X, Qiu J. Recent advances in energy transfer in bulk and nanoscale luminescent materials: from spectroscopy to applications. Chem Soc Rev. 2015;44:8714–46. https://doi.org/10.1039/x0xx00000x.
4. Jones RR, Hooper DC, Zhang L, Wolverson D, Valev VK. Raman techniques: fundamentals and frontiers. Nanoscale Res Lett. 2019;14(1):34. https://doi.org/10.1186/s11671-019-3039-2.
5. Butt MA, Kazanskiy NL, Khonina SN, Voronkov GS, Grakhova EP, Kutluyarov RV. A review on photonic sensing technologies: status and outlook. Biosensors. 2023;13(5):1–33.
6. Lundstrom M, Ren Z. Essential physics of carrier transport in nanoscale MOSFETs. IEEE Trans Electron Devices. 2002;49(1):133–41.
7. Das A, Kanaujia BK, Deswal SS, Rewari S, Gupta RS. Doping induced threshold voltage and I ON/I OFF ratio modulation in surrounding gate MOSFET for analog applications. In: IEEE international conference of electron devices society kolkata chapter (EDKCON). 2022. pp. 1–6. https://doi.org/10.1109/EDKCON56221.2022.10032940.
8. Das A, Kanaujia BK, Nath V, Rewari S, Gupta RS. Impact of reverse gate oxide stacking on gate all around tunnel FET for high frequency analog and RF applications. In: 2020 IEEE 17th India council international conference, INDICON 2020. 2020. pp. 1–6.
9. Goyal P, Srivastava G, Madan J, Pandey R, Gupta RS. Analytical modelling and reliability analysis of charge plasma-assisted $Mg_2Si/Si$ heterojunction doping less DGTFET for low-power switching applications. Phys Scr. 2024;99(1): 015008. https://doi.org/10.1088/1402-4896/ad1232.
10. Goyal P, Srivastava G, Madan J, Pandey R, Gupta RS. A $Mg_2Si/Si$ heterojunction based dielectric modulated dopingless TFET biosensor for label free detection. Mater Sci Eng, B. 2024;304: 117356. https://doi.org/10.1016/j.mseb.2024.117356.
11. Goyal P, Srivastava G, Madan J, Pandey R, Gupta RS. Source material valuation of charge plasma based DG-TFET for RFIC applications. Semicond Sci Technol. 2022;37(9): 095023.
12. Bhardwaj A, Kumar P, Raj B, Anand S. Design and performance optimization of dopingless vertical nanowire TFET using gate stacking technique. J Electron Mater. 2022;51(7):4005–13. https://doi.org/10.1007/s11664-022-09658-9.
13. Bhardwaj A, Kumar P, Raj B, Kumar N, Anand S. Design and optimization of vertical nanowire tunnel FET with electrostatic doping. Eng Res Express. 2023;5(4): 045025. https://doi.org/10.1088/2631-8695/acff3a.
14. Su YS, Chang HC. Bridging photonics and optoelectronics curriculum for the solar photovoltaic and LED industries. Int J Photoenergy. 2014;2014(1): 146810. https://doi.org/10.1155/2014/146810.
15. Li N, Ho CP, Xue J, Lim LW, Chen G, Fu YH, et al. A progress review on solid-state LiDAR and nanophotonics-based LiDAR sensors. Laser Photonics Rev. 2022;16(11):1–24.
16. Camposeo A, Persano L, Farsari M, Pisignano D. Additive manufacturing: applications and directions in photonics and optoelectronics. Adv Opt Mater. 2019;7(1):1800419.
17. Butt MA, Mateos X, Piramidowicz R. Photonics sensors: a perspective on current advancements, emerging challenges, and potential solutions (Invited). Phys Lett. 2024;516: 129633. https://doi.org/10.1016/j.physleta.2024.129633.
18. Das A, Rewari S, Kanaujia BK, Gupta RS. Recent technological advancement in surrounding gate MOSFET for biosensing applications—a synoptic study. SILICON. 2022;14(10):5133–43.
19. Das A, Rewari S, Kanaujia BK, Deswal SS, Gupta RS. Analytical modeling and doping optimization for enhanced analog performance in a Ge/Si interfaced nanowire MOSFET. Phys Scr. 2023;98(7):74005. https://doi.org/10.1088/1402-4896/acde16.
20. Das A, Rewari S, Kanaujia BK, Deswal SS, Gupta RS. Ge/Si interfaced label free nanowire BIOFET for biomolecules detection-analytical analysis. Microelectron J. 2023;138:105832. https://doi.org/10.1016/j.mejo.2023.105832.
21. Yadav S, Das A, Rewari S. Dielectrically-modulated GANFET biosensor for label-free detection of DNA and Avian influenza virus: proposal and modeling. ECS J Solid State Sci Technol. 2024;13(4): 047001. https://doi.org/10.1149/2162-8777/ad3364.

22. Das A, Rewari S, Kanaujia BK, Deswal SS, Gupta RS. Numerical modeling of a dielectric modulated surrounding-triple-gate germanium-source MOSFET (DM-STGGS-MOSFET)-based biosensor. J Comput Electron. 2023;22(2):742–59. https://doi.org/10.1007/s10825-023-02008-w.
23. Das A, Rewari S, Kanaujia BK, Deswal SS, Gupta RS. Analytical investigation of a triple surrounding gate germanium source metal–oxide–semiconductor field-effect transistor with step graded channel for biosensing applications. Int J Numer Model Electron Networks Devices Fields. 2023;36(6): e3106. https://doi.org/10.1002/jnm.3106.
24. Das A, Rewari S, Kanaujia BK, Deswal SS, Gupta RS. Physics based numerical model of a nanoscale dielectric modulated step graded germanium source biotube FET sensor: modelling and simulation. Phys Scr. 2023;98(11): 115013. https://doi.org/10.1088/1402-4896/acf4c9.
25. Das A, Rewari S, Kanaujia BK, Deswal SS, Gupta RS. Analytical characterization of a label free Si/InAs hetero-interfaced cylindrical BioFETD for biosensing applications. Micro and Nanostruct. 2025:208152. https://doi.org/10.1016/j.micrna.2025.208152.
26. Yariv A, Yeh P. Photonics: optical electronics in modern communications. New York: Oxford University Press; 2007.
27. Sharma S, Tharani L. Photonics for AI and AI for photonics integration: materials and characteristics. J Inf Optim Sci. 2024;45(3):805–16.
28. Poonguzhali S, Sivasangari A, Ajitha P, Lalithakumari S, Sridevi A, Danasegaran SK. Design and performance analysis of smart photonic sensors for industrial applications. Curr Appl Phys. 2022;39:183–9. https://doi.org/10.1016/J.CAP.2022.04.006.
29. Aloupogianni E, Doctor F, Karyotis C, Maniak T, Tang R, Iqbal R. An AI-based digital twin framework for intelligent traffic management in Singapore. In: International conference on electrical, computer, and energy technologies, ICECET 2024. 2024. pp. 1–6. https://doi.org/10.1109/ICECET61485.2024.10698642.
30. Seas A, Robinson B, Shih T, Khatri F, Brumfield M. Optical communications systems for NASA's human space flight missions. In: International conference on space optics—ICSO 2018. 2019. pp. 182–91. https://doi.org/10.1117/12.2535936.
31. Wu Y, Liu X, Pandey A, Zhou P, Dong WJ, Wang P, et al. III-nitride nanostructures: emerging applications for Micro-LEDs, ultraviolet photonics, quantum optoelectronics, and artificial photosynthesis. Prog Quantum Electron. 2022;85: 100401. https://doi.org/10.1016/j.pquant elec.

# Chapter 4
# Integration of Photonics and Optoelectronics in Industry 5.0

**Soumya Rai, Chhaya Ravi Kant, and Ranjana Jha**

**Abstract** The chapter emphasizes the significance of photonics and optoelectronics for enabling Industry 5.0, a new era when cutting-edge technologies are seamlessly coupled with human-oriented ways in all sorts of industrial processes. Photonics, the science of using light to move information and other objects in ways similar to how electricity is used electronically, along with optoelectronics or electronics combined with optical functionality is transforming multiple industrial fields. Discussing high-tech use cases, it covers their application in smart manufacturing to improve accuracy and time to market; advanced communication systems to increase data speed at reduced costs as well examples of cutting-edge R&D concerning optoelectronic sensors for next-generation sensor web-based process systems. The chapter also discusses synergies between photonics, artificial intelligence, and the Internet of Things enabling intelligent and adaptive industrial systems. This integration will allow Industry 5.0 to design industrial environments that are both smarter, and more sustainable as well as human-friendly. This chapter presents an all-inclusive account of the recent progressions, obstacles, and prospective landscapes in photonics and optoelectronics conforming to their disruptive influence on industrial arena.

**Keywords** Photonics · Optoelectronics · Smart manufacturing · IoT · AI · Advanced sensors

## 4.1 Introduction

Industry 5.0 is the next phase in the evolution of the industry that combines more human-centric approaches with advanced technologies including artificial intelligence (AI), robotics, the Internet of Things (IoT), and machine learning [1–3]. These stimuli also lead to increasingly agile, cost-effective, and resilient industrial ecosystems. While Industry 4.0 focused mainly on automation and smart factories [4],

S. Rai (✉) · C. R. Kant · R. Jha
Department of Applied Sciences and Humanities, Indira Gandhi Delhi Technical University for Women, Delhi, India
e-mail: soumyarai.sr16@gmail.com



V. Bhardwaj et al. (eds.), *Photonics and Optoelectronics in Industry 5.0*, Advances in Optics and Optoelectronics, https://doi.org/10.1007/978-981-96-7594-4_4

Industry 5.0 considers the symbiosis of machines and humans working together [5, 6], supercharging productivity while maintaining sustainable practices by ensuring satisfaction of workers. Photonics and optoelectronics are the two foundational technologies at the core of this reengineering.

Photonics, the control of light as information and in terms of interacting with atoms or bulk matter, has long been a subject of interest beyond simple information technology and computing solutions [7]. It is increasingly seen as vital for fostering innovation in other industries. With the help of photonics, light can be precisely controlled in areas such as material processing, sensing, communication, and data storage, thereby enabling the realization of Industry 5.0. Lasers enable manufacturing processes such as precision cutting, welding, and material deposition [8]. Photonics integration into industrial processes improves speed, accuracy, and cost efficiency. Photonics is also essential in communication systems, which operate typically over fiber optics [9], with higher bandwidth and data rates compared to traditional copper wires. In a world that sees vast amounts of data being produced by IoT sensors and AI systems which require sending and receiving at speeds, this type of functionality is without doubt increasingly important. In addition, technological progress in the control of light at nanoscale and quantum systems is ambushing computing by developing photonic devices, which are expected to be superior to current electronic systems.

Optoelectronics works at the intersection of optical and electronic technologies, dealing with the devices that convert light or electrical signals into each other [10]. This enables various applications from telecommunications and health care to manufacturing. Optoelectronic devices represent the heart of today's communication networks, enabling data transfer at high speed with very low energy consumption, and are dominated by key components such as LEDs, photodetectors, and laser diodes. In Industry 5.0, communication systems [11] are optimized with optoelectronics especially due to the emergence of 5G and next-generation networks capable of transferring large datasets that are essential for industries requiring real-time data processing and remote monitoring. Furthermore, optoelectronic sensors also provide precise identification of crucial features such as temperature, pressure, and chemical composition required for operational effectiveness and security. In order to minimize the environmental burdens, industries are trying to design various optoelectronic devices with lower energy consumption which exceed high-gain performances. For example, solar cells convert light directly into electrical energy, thus contributing to the use of sustainable energy instead of traditional energy sources. Continued advancement in these technologies could help develop a greener industry for Industry 5.0 processes that would allow factories operate toward copulsing our environmental targets. Table 4.1 lists the various applications of photonics and optoelectronics in multiple sectors of Industry 5.0. The future industrial space will experience exponential efficiency, better sustainability, and a higher quality of life for workers enabled by the strong synergy between photonics and optoelectronics—the next big step in the evolution of industries.

**Table 4.1** Integration of photonics and optoelectronics across various sectors of Industry 5.0

| Industrial sector | Integration of photonics | Integration of optoelectronics |
|---|---|---|
| Smart cities | Intelligent lightning sensors for smart infrastructure | IoT devices for smart lighting, data transmission |
| Manufacturing | Laser cutting, welding, material processing | LED lighting, photodetectors for quality control |
| Telecommunications | Optical fiber communications for high-speed data transfer | Fiber optic sensors, laser diodes for signal transmission |
| Health and medicine | Photonic imaging using MRI, CT scan for diagnostics | Biosensors and optoelectronic diagnostic tools |
| Energy and environment | Solar cells, energy-efficient lighting | Photovoltaics, light-based energy harvesting systems |
| Automotives | LiDAR for autonomous vehicles | Optoelectronic sensors for vehicle safety and efficiency |

## 4.2 Impact of Photonics and Optoelectronics in Shaping Industry 5.0

Photonics and optoelectronics are the underlying technologies that drive Industry 5.0 forward. Their complementary strengths in the manipulation of light and the functional integration of optics into electronic systems have opened an entirely new chapter in industrial advancement, one that goes beyond mere automation to realize smart, sustainable, and human-centric industrial spaces. This section focuses on their basic principles and tasks.

### *4.2.1 Basics of Photonics: Using Light for Information and Process Control*

Photonics, the science of light, deals with processes concerning generation, manipulation, and detection of light particles. In industrial use, light is not only used for illumination but also as an information transmitter and process regulator [12]. Controlling light with precision is essential for the full range of applications such as material processing, high-resolution imaging, and data communication. Lasers are exemplified in a lot of photonic technology, yet they are the commonplace tools in the industry, used for thin cutting with high precision, or welding and engraving. In industries such as aerospace, automotive, and electronics, their potential to focus light in a single beam provides accurate level of precision which traditional forms of manufacturing are unable to. In addition, advancements in photonic technologies such as optical sensors help to control production processes based on real-time monitoring which improves product quality and eliminates waste. These have already been installed in smart factories, for example to measure temperature, pressure, and chemical composition with completely noninvasive photonic systems ensuring smooth functioning. Communication is another field where photonics has a very strong impact as well. Optical fibers are one of the most important pieces of modern telecommunications infrastructure that have become essential for high-speed, high-bandwidth communication through faster light transmission than conventional copper wiring. The importance of photonic communication for passing real-time data between IoT and smart network systems continues to increase especially in light of the evolution to data-driven industries.

### 4.2.2 *Basics of Optoelectronics: Integration of Optical and Electronic Technologies*

Optoelectronics is a subcategory of electronics that integrates both electronic and optical systems to perform sensing (light detection), communication (emission as well as modulation), or micro-processing tasks [13]. The unique advantage of optoelectronic devices is that they provide the speed and bandwidth of light combined with the precision and control of electronics. This includes a range of parts and systems that are needed for sensing information, communication, as well as energy effectiveness. Optoelectronics is expected to have a crucial influence in the formation of Industry 5.0 with intelligent opto-sensing response ecosystem. This means that optoelectronic devices like light-emitting diodes (LEDs), photodiodes, and laser diodes are already employed in industrial settings to save energy and support communication. Photodetectors, for instance, are an essential component in automation systems since they can detect changes in light levels and prompt responses in real-time manufacturing plants. Sensors such as optoelectronic sensors are used for the monitoring of several parameters including structural health monitoring in buildings and to detect chemical changes happening in smart manufacturing systems. Additionally, it facilitates developing communication technologies such as fiber optics—a key to the high-speed real-time data transmission for industrial IoT systems. Optical interconnections via the collaboration of optoelectronics and photonics enable improved and reliable communication that will be required in large data volume processing and exchange use cases. Figure 4.1 depicts the various photonics and optoelectronic devices based on 2D materials.

## 4.3 Applications of Photonics and Optoelectronics in Smart Manufacturing

The pillar of smart manufacturing embodies the spirit of Industry 5.0 [14] through advancements in photonics and optoelectronics, including their convergence, highly defined research directions in efficient use of photons, development of optical

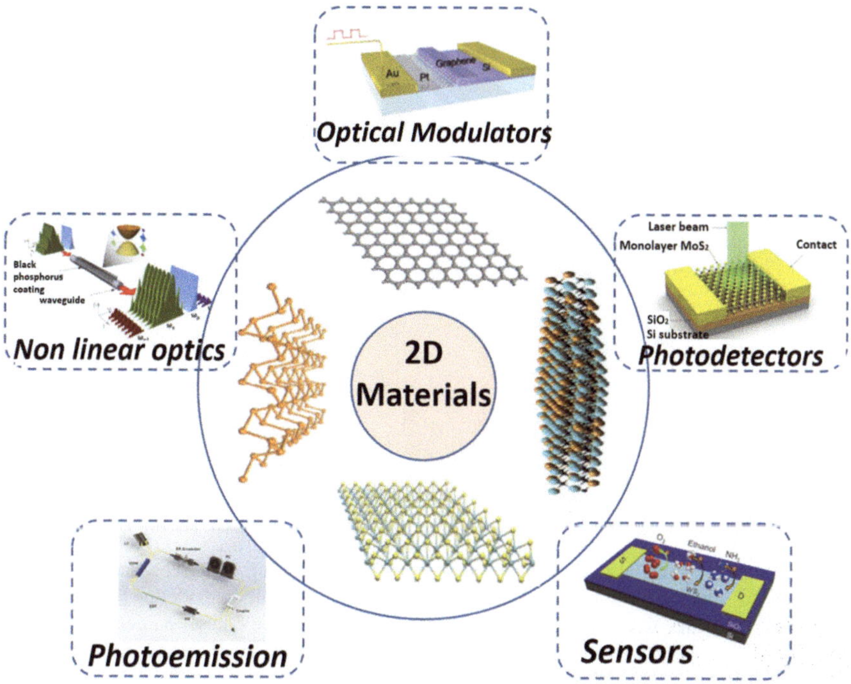


**Fig. 4.1** Various photonics and optoelectronic devices based on 2D materials reproduced from [13] Copyright ©2025 AIP Publishing LLC. Used under the Creative Commons Attribution (CC-BY) license

instrumentation, and ultrafast laser applications. This section details the impact of photonics and optoelectronics on the manufacturing sequence and the opportunities in achieving smarter and greener production systems.

### *4.3.1 Precision Material Processing with Photonic Technologies*

Material processing has been transformed with the aid of photonics, specifically lasers which offer control, high precision, and flexibility. Technologies based on lasers are being commonly used in various sectors such as automotive, aerospace, and electronics for welding and cutting processes of materials. Manufacturers are able to effectively cut materials with minimum damage and wastage using the ability of the laser beams to focus into narrow and intense beams. The structural integrity of the high-performance machineries including aircrafts and automobiles is maintained due to the strong joints made by lasers during the welding process [15] as illustrated in Fig. 4.2a. Additive manufacturing, commonly recognized as 3D printing, builds

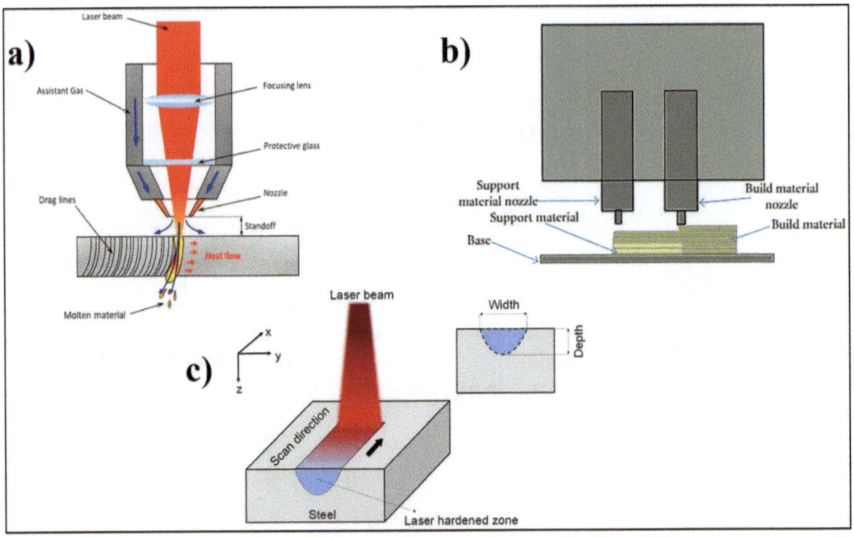


**Fig. 4.2** Material processing techniques using lasers **a** laser cutting setup reproduced from [15]  **b** fused deposition modeling for 3D printing reproduced from [16]  **c** Laser hardening technique reproduced from [17] 

objects employing photonics technologies like laser melting and laser sintering in a layer-by-layer process. One such example of a widely used technique is fused deposition modeling [16] as depicted in Fig. 4.2b. This technique permits the fabrication of customized, sophisticated components which would be tough or nearly impossible to fabricate with general manufacturing procedures. Using photonics, design freedom is increased along with minimizing wastage of material, resulting in further cost savings and environmental benefits. Lasers are also employed for surface treatments, such as coating, hardening, or texturing materials as shown in Fig. 4.2c [17]. These methods enhance the wear resistance, functionality, and durability of products, specifically in industries where materials are exposed to harsh conditions, including energy sectors and aerospace.

### *4.3.2 Optoelectronic Sensors for Real-Time Monitoring and Quality Control*

Optoelectronics boosts smart manufacturing by providing advanced sensing and monitoring capabilities which ensure high-quality production while minimizing downtime and wastage. Optoelectronic sensors like photodetectors [18] are used for multiple parameters including real-time checks for pressure, temperature, chemical composition, and light intensity. In many industries, especially those of pharmaceuticals, food processing, electronics, etc., these sensors play an essential role in ensuring the quality and consistency of the products. Microscopic changes in the properties of materials as they are created could be detected by the photonic sensors allowing rapid adjustments to maximize uniformity of the final product. The data generated by optoelectronic sensors can be used in feedback loops integrated with automation systems that autonomously modify manufacturing parameters. Without this adaptability, it becomes impossible to realize the future of industrial systems envisioned by Industry 5.0, intelligent, self-optimizing approaches. These feedback loops minimize human intervention, increase production speed, and help improve the quality of the product in smart factories. Optoelectronic devices pave the way for nondestructive testing methods [19], a specific type of inspection that does not cause material or component damage and therefore is very valuable to manufacturers. This is particularly critical in industries such as aerospace, automotive, and medical device manufacturing, where safety and reliability are paramount. Infrared thermography, X-ray imaging, and laser ultrasound are in short examples of optical techniques that can be used to inspect the integrity of materials to detect defects, cracks, and deformations and as such ensure only high-quality products make it into the market.

### *4.3.3 Advanced Communication Systems*

Data transmission systems have experienced a plethora of advantages owing to characteristics of photonics and optoelectronics in terms of speed, reliability, and energy efficiency. In this section, we describe the capabilities of photonic and optoelectronic technology that facilitate communication infrastructure and emerging applications including combinations of cloud computing and APNs, and the potential impact on future industrial communication networks.

(a) **Photonics in High-Speed Data Transmission**

Fiber optic communication systems have revolutionized data transmission with the help of photonics. Since light can travel far with least loss and at extremely high speeds, photonic systems utilize light as the means for information transmission. Fiber optics now serve as the spinal column of global communications networks including everything from on the internet to telecommunications and industrials communications. Fiber optic cables transmit data over long distances using light signals, far outperforming older copper wires [9]. This is accomplished by total internal reflection, allowing light signals to bounce the length of a fiber's core without much loss. They find use in areas where wide bandwidth and low attenuation are required, making fiber optics a common medium for high-speed customer data services, long distance communication, and industrial data networks. Figure 4.3 depicts the structural components of a fiber optic cable. It is a technology used to allow multiple data signals to be sent simultaneously over a single optical

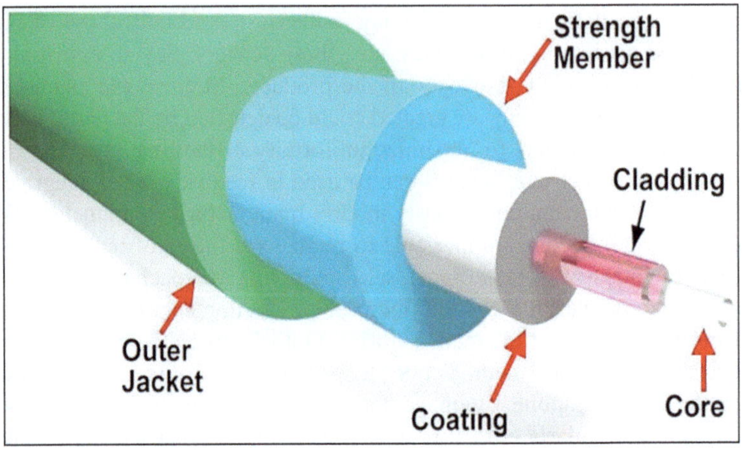


**Fig. 4.3** Structural components of a fiber optic cable used for optical communication and data transfer reproduced from [12] Copyright © 1996–2025 MDPI (Basel, Switzerland). Used under the Creative Commons Attribution (CC-BY) license

fiber cable by using light of different wavelengths [20]. This significantly expands the data-carrying capacity of communications networks, enabling more information to be transmitted with less infrastructure and higher speed and efficiency. Photonic integrated circuits (PICs) integrate multiple on-chip photonic components, including lasers, modulators, and detectors, in high-performance optical data transmission formats that are small and energy efficient [21]. PICs are a key enabler for data centers, telecommunication and industrial communication system evolution, enabling faster data processing and transmission while also minimizing power and footprint requirements.

(b) **Optoelectronic Devices for Communication Systems**

Optoelectronic devices are a crucial part of modern communication systems, bridging the gap between electronic components and optical signals such as light-emitting diodes (LEDs), photodetectors, lasers, and modulators. Such devices are important not only to fiber optic communication but also for wireless communication technologies that are speedily expanding in Industry 5.0. Some of these devices are shown in a flowchart depicted in Fig. 4.4. Light-emitting diodes (LEDs) and lasers are the main light sources used in optical communication systems to transmit data. Lasers provide high coherence and intensity [22] and hence are the best candidates for communication systems with long distance and high capacity. Due to their cheap cost and energy-efficient nature, lit LEDs are mainly used in low range applications. In optical communication systems, signals that are initially carried in the form of light must be detected so they can be converted back into the electrical signals used by computers—and this is where photodetectors come into play. Such devices including PIN diodes and avalanche photodiodes (APDs) are crucial for the speed and the accuracy of data transmission [23]. Photodetectors are also necessary in wireless communication systems since they are responsible for the detection and decoding of transmitted light using free-space optics (FSO) as well as other wireless media. In addition, they can be used in modulating light signal properties, such as amplitude, phase, and frequency, which will make the accurate inclusion of data into optical carriers possible. Electro-optic modulators, such as lithium niobate modulators, are the ones that are widely used in the high-speed communication systems. They allow for efficient data transfers over fiber optic networks [24].

(b) **Future of Industrial Communication: Photonics, 5G, and Beyond**

The integration of photonics and optoelectronics with the next-generation technology that includes 5G and 6G will bring the faster, more reliable, and intelligent industrial communication systems, which will thus make a way for Industry 5.0. Photonics will

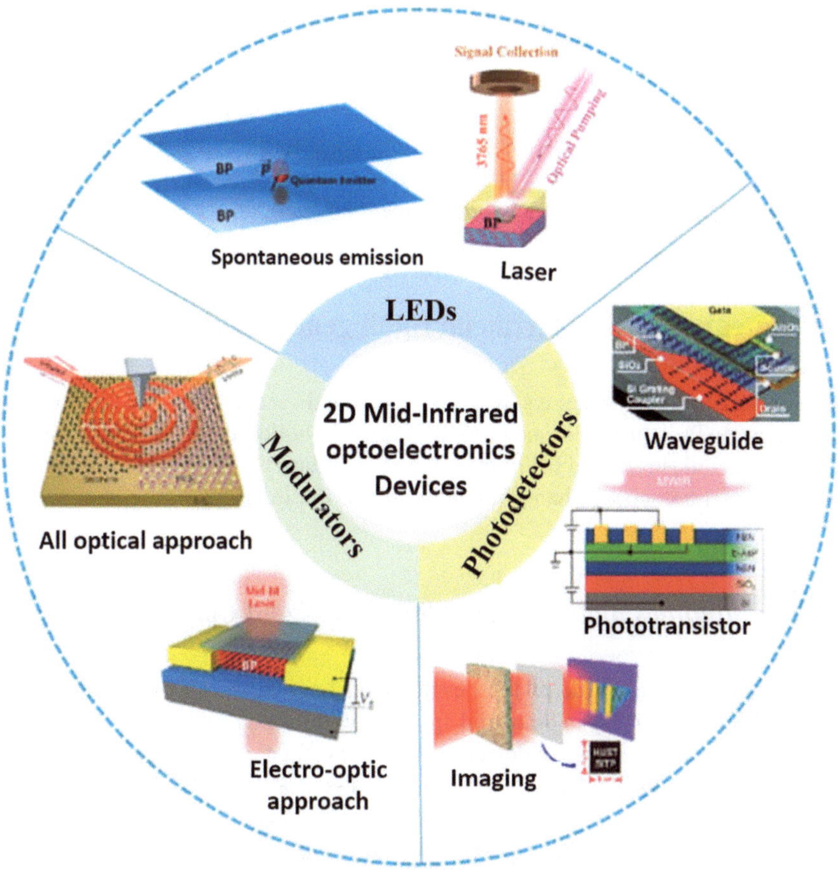


**Fig. 4.4** Various optoelectronic devices based on 2D materials for communication and imaging systems reproduced from [25] Copyright © 1996–2025 MDPI (Basel, Switzerland). Used under the Creative Commons Attribution (CC-BY) license

be a power horse during 5G network expansion [26], and the networks will promise to bring more speed and less latency in wireless communication. Optic fibers are going to be the basis of the high data rates required by 5G, and optoelectronic devices will be employed to interface between wireless and fiber optic networks. Consequently, this will ensure smooth and efficient communication between machines, sensors, and humans of the smart factories, thus increasing productivity and enabling real-time decision-making. Free-space optical communication (FSO) is a wireless communication technology that makes use of light to send data in the air, instead of through fiber optic cables [27]. FSO is slowly turning to be a feasible solution in places where physical cables cannot be laid, not to mention where the cost is skyrocketing compared to the benefits involved. In the industrial field, FSO can be employed to establish high-speed data links between machines, sensors, and control systems, thus supporting the wireless connectivity which is required by Industry 5.0. Over and above 5G, the building of 6G networks (Fig. 4.5) will not only rely on photonic and optoelectronic technologies [28], but it will be photonic and optoelectronic technologies that will lead to the developments required for its complete and efficient implementation. 6G has the potential to achieve the data transmission speed which is way higher than 5G and will have low latency and higher energy efficiency. Photonic technologies should be responsive to the ultrafast data traffic that 6G will be requiring and optoelectronic devices will facilitate the development of AI, IoT, and machine learning systems in the industrial communication networks.

## 4.4 Integration with Artificial Intelligence and IoT

The convergence of photonics with artificial intelligence (AI) and the Internet of Things (IoT) is providing new opportunities to build adaptive, intelligent, and autonomous industrial systems in the context of Industry 5.0. This section explores how these technologies go hand-in-hand to transform industries.

### *4.4.1 Photonics and IoT: Enabling Intelligent Connectivity*

Photonics takes a lead role in the development and expansion of the Internet of Things (IoT), particularly in industrial applications [30]. IoT requires the continuous connection between billions of devices and sensors that are capable of real-time data collection and exchange. Photonics is gaining attention in IoT due to the capabilities of high-speed and low-latency transmission of data and real-time monitoring, control, and decision-making in the industrial environment [31]. High-speed data transmission in IoT is only possible by photonics. IoT can only succeed in managing massive amounts of interconnected devices having data generated on an everyday basis if photonics is integrated with it [32]. Fiber optic communication systems and photonic sensors are the infrastructure used for data transmission between IoT

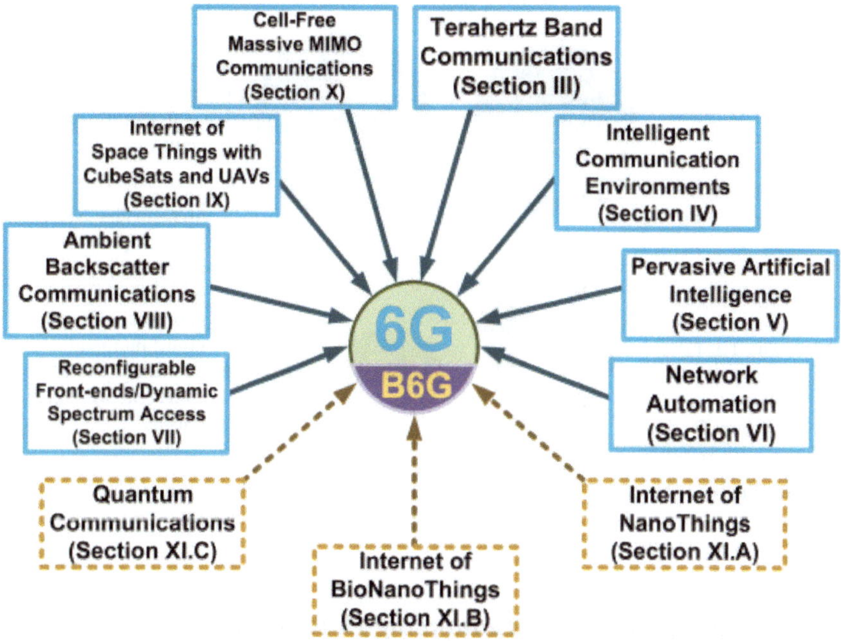


**Fig. 4.5** Future technologies for enabling 6G network and beyond reproduced from [29] Copyright © 2020 IEEE. Used under the Creative Commons Attribution (CC-BY) license

devices [9]. This advancement allows industries to implement real-time monitoring and control systems that provide reliable operation, lower downtime, improvements in decision-making processes, and the use of integrated data from multiple suppliers in real time. Besides, the photonics-enabled IoT sensors (depicted in Fig. 4.6) are very advantageous in IoT applications because they can sense a variety of information with very high precision and speed [33]. There are some benefits that come with them; for instance, when the IoT network is set up, photonics sensors help make decisions in real time, reduce production disruptions, and maintain product quality control. Moreover, the IoT technology combined with photonics offers the introduction of smart manufacturing systems where machines, sensors, and production lines are interconnected and communicate with each other. The data that are sent and received during the manufacturing process in the form of real-time communication are facilitated by the technology of photonic communication systems and sensors. This is an important component in the manufacturing process since it allows the automation of decision-making, predictive maintenance, and full flexibility in production. This leads to the optimization of the manufacturing process with the result of a higher level of customization and reduction of waste.

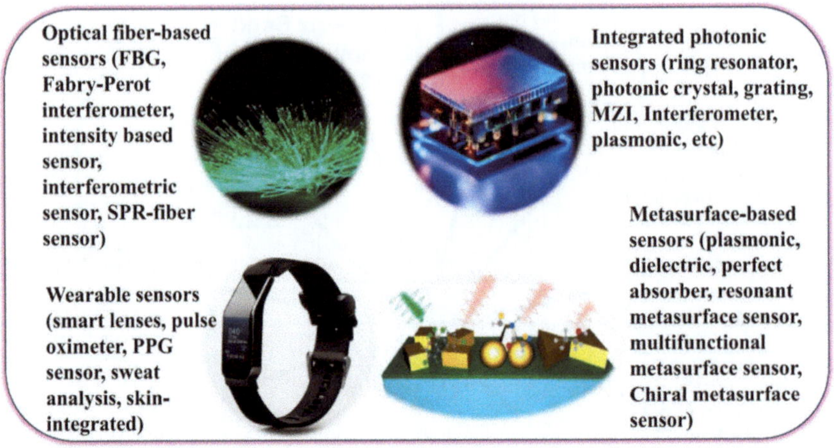


**Fig. 4.6** Photonics-based IoT sensors for specific applications and platforms reproduced from [33] 

### *4.4.2 AI and Photonics: Enhancing Data Analysis and Decision-Making*

Photonics is complemented by artificial intelligence (AI) through more sophisticated data analytics and the capability of decision-making [34]. As soon as the given data are sensed and transmitted by using photonics techniques, AI leads to data analysis by generating solutions and making the systems smarter and more flexible [35]. The newly formed cooperation of AI and photonics will allow for the development of greater efficiency in the circuits as well as the improvement of the quality of the products and the cost of running. The AI algorithms are capable to process the huge amount of data the photonic and communication systems come with, detect the occurring patterns, foresee the trends and taking the forthwith decision based on the real-time data. Nevertheless, on the other hand, in the factories, AI-based photonic systems can be used to produce the optimized lines, monitor and reduce the consumption of energy, and can assure the quality control by the identification of the defects or inefficiencies in real time.

Machine learning, which is a subset of AI, is among the most important point of the conversation between AI and photonics. These algorithms are capable of prophecies such as the possible equipment failure, necessary parameters optimization, and even the capability to the change the production requirements [36] through the analysis of the historical data and the learning from the previous incidents. Various studies are published on the use of AI and machine learning as seen in their increasing graph in Fig. 4.7. Photonic sensors and communication systems which are the main devices collecting and transmitting data necessary for the functioning of the intelligent systems aid the industrial processes to be more efficient and adjustable over

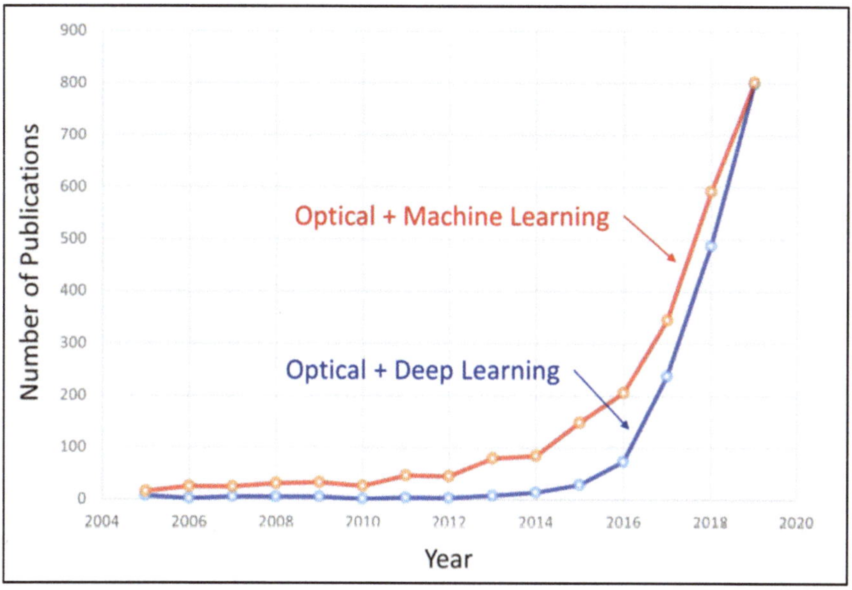


**Fig. 4.7** Graph depicting the increase in the number of publications based on AI and machine learning reproduced from [37] Copyright © 2025 AIP Publishing LLC. Used under the Creative Commons Attribution (CC-BY) license

time. Predictive maintenance is one of the key applications of AI's integration with photonics. Starting with an AI-designed space where the photonics-based system is proactive and self-sustaining helps industries achieve in time production with minimal wear, tear, or potential failures. In the context of environmental sustainability, it is predictable that applications such as additive manufacturing, the digitalized service economy, and the electric vehicle economy are points of focus; yet, they are only the beginning of the process of change.

### *4.4.3 AI, IoT, and Photonics for Adaptive Industrial Systems*

The integration of photonics, AI, and IoT leads to the development of adaptive industrial systems that can immediately adjust production requirements and equipment conditions [38]. These adaptive systems are able to combine all the three technologies; thus, industries can have more flexibility, efficiency, and resilience during their operations. The synergy of photonics, AI, IoT, and adaptive systems' self-optimization based on real-time data are the main driving forces behind this. For instance, photonic sensors in a manufacturing environment could be used to analyze production variables, while the AI algorithms would process the data and regulate the process parameters to maintain the optimal performance. The idea of IoT devices

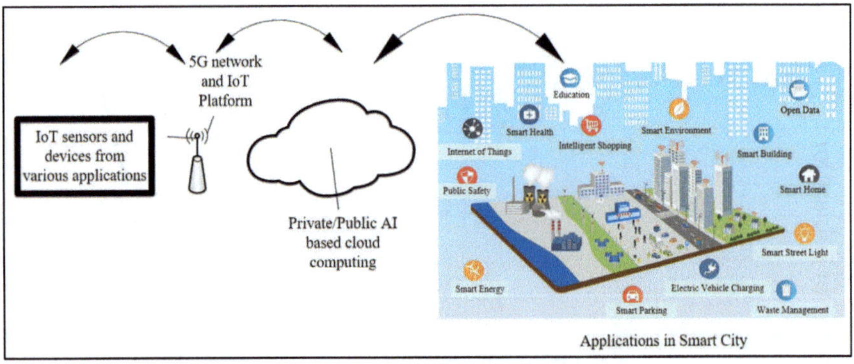


**Fig. 4.8** A visual representation of how AI, IoT, and 5G network work together to form a futuristic smart city reproduced from [39] Copyright © 1996–2025 MDPI (Basel, Switzerland). Used under the Creative Commons Attribution (CC-BY) license

in this case means continuous communication between machines and sensors by creating a feedback loop that allows them to always be at the top of their game. Smart factories are the place where photonics, AI, and IoT become one; thus, real-time decision-making is possible, and machines and sensors collaborate to dynamically adjust the manufacturing processes with changing demand. Consequently, more production customization is achievable together with optimal efficiency and minimal waste. Smart factories are an integral part of future technology—Industry 5.0, which is the industrial systems that are designed to be highly adaptable and responsive to market demands. The cooperation among these technologies also aids innovation along with sustainability in context with the development of industrial systems. Photonic sensors scan environmental conditions and energy consumption, while AI algorithms regulate resource usage and save energy to the maximum extent. IoT communication using devices that ensure all parts of the system are interconnected and responsive to the conditions that change is another buffer that industrial systems benefit from. As a result, industrial systems can withstand disruptions and are better adapted to the sustainable use of resources and well aligned with the goals of Industry 5.0 as depicted in Fig. 4.8.

## 4.5 Future Prospects of Photonics, AI, and IoT Integration and Challenges

The future of industrial systems is chiefly dependent on the integration of photonics, AI, and IoT. The desires for the innovation of data transmission speed, sensor accuracy, and energy efficiency have ushered in a new era of photonic technologies. AI will further humanize machines, with enhanced algorithms in predicting maintenance, optimization, and real-time analysis. IoT will be the issue of the future with more devices, sensors, and machines contributing to the ever-growing interconnected industrial ecosystem. Relating to future development is the exploitation of quantum photonics creation of devices that obey the laws of quantum mechanics to communicate and have a better hearing. Quantum photonics can incredibly change AI and IoT systems by creating an environment where communication is done in an unbelievably secure way and data processing has been much improved [40]. As much as quantum technologies get developed, they will also help photonics to be further integrated with AI and IoT in industrial systems. AI-driven photonic networks are another area of future research [41]. These networks will use AI to build and manage photonic communication systems in real time, thus being in the best position to save time and reduce the cost of data transmission systems. AI enhancing photonic networks will be the best bet for the industry to achieve higher automation and adaptability. The intertwining of photonics, AI, and IoT will contribute to the establishment of sustainable industrial undertakings. These are technologies that will be utilized in the transition of industries to more energy-efficient and less wasteful and also in the maintenance of environmental crises. Photonic sensors, along with AI-generated optimization, would assist companies in reducing their use of energy, the generation of waste, and the emission of carbon dioxide.

The integration of photonics and optoelectronics into Industry 5.0 presents great potential, but also showcases fair share of challenges. This section explores some of the major challenges, ongoing research in optoelectronic sensors for next-generation industrial systems, and future opportunities for developing sustainable, human-friendly industrial environments.

### *4.5.1 Key Technical Challenges in Widespread Adoption*

Despite the transformative potential of photonics and optoelectronics, various technical challenges hinder their seamless integration into Industry 5.0 as illustrated by the flowchart in Fig. 4.9. These issues require immediate solution of the full potential of these technologies in the industrially operated environment [42]. The first and the foremost issue is the huge cost of the deployment of the photonics and optoelectronics systems. These technologies provide multiple benefits in terms of efficiency and sustainability, but the initial investment for installation, calibration, and maintenance can be major. In most sectors, especially in small and medium-sized enterprises (SMEs), the cost of entry to this kind of technology is the main reason for not going. Photonic devices, sensors, and manufacturing processes are expensive, and thus, lowering their expenses will be the only way for their wider uptake. The introduction of photonics and optoelectronics into the current industrial system at times needs to be done by experts. Mostly, they require a lot of specialized knowledge and precise calibration for their proper functioning in different industrial environments. Additionally, the seamless interoperability among photonic systems and other digital technologies like AI and IoT will be more challenging [43]. Overcoming the integration impediments will be possible by the progress in the area of

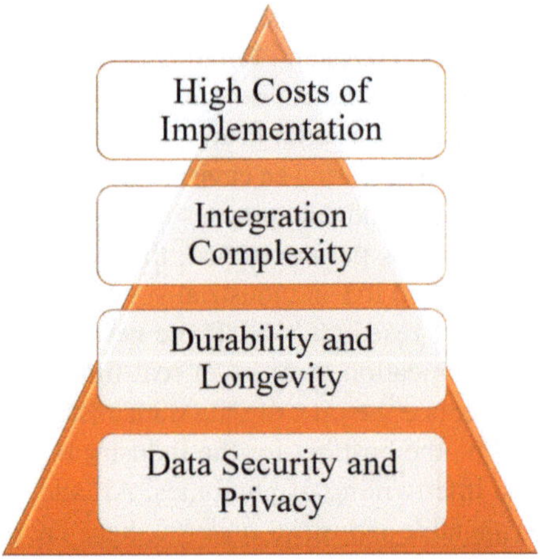


**Fig. 4.9** Major challenges in integration of photonics and optoelectronics in Industry 5.0

user-friendly interfaces, the standardization of photonic components, as well as, by the improvement of system compatibility. One must guarantee that these photonic and optoelectronic mechanical systems can transmit data in the correct way and perform as planned for the required service. Being precise and effective are the main characteristics of devices that work by changing the information from light to other forms. Their endurance and functioning capacity under severe environmental conditions still pose one of the crucial issues. That is, temperatures, for example, solar radiation, toxins, corrosion, and mechanical loads can deteriorate the operation of these devices over time. The development of tough and hard-wearing photonic materials and parts that are capable of withstanding the tough conditions in the industry will be an important long-term step. Along with AI, IoT, and other digital technologies, as they become part of photonic and optoelectronic systems, the area of data security gains a primary concern. Photonic sensors and the devices that are optoelectronic deal with big volumes of secret information, and they are mostly responsible for the collection and transmission of this information. The major step to be followed by the industries which aim to adopt these technologies on a wide scale is to keep these data safe from external cyberattacks and breaches. Therefore, to counter this challenge effectively, enhanced encryption methods and safe protocols of transmission of data are highly crucial.

### *4.5.2 Emerging Research in Optoelectronic Sensors for Next-Generation Industrial Systems*

Optoelectronic sensors are trending in the research field for the next-generation industrial systems, with the ability to revolutionize Industry 5.0 by providing smarter, more adaptive, and human-friendly environments. Recent efforts in research are aimed to address the challenges discussed above while pushing the boundaries of what these sensors can achieve.

(a) **Next-Generation Optoelectronic Sensors**: Researchers are currently focused on the design of optoelectronic sensors being more sensitive, faster responses, and more reliability. Nevertheless, industrial processes [44] conditionally say these technologies are indispensable for real-time monitoring and control. Hence, optoelectronic sensors will soon be capable of measuring small changes in temperature and pressure that can predict maintenance, provide accurate production, and increase safety in industrial areas.

(b) **Flexible and Wearable Optoelectronic Sensors**: The ongoing work on flexible and wearable optoelectronic sensors integrated into the human-centric environments of the Industry 5.0 is an additional field of research. It has to be emphasized that these sensors can be worn as clothes and be attached to the body providing continuous monitoring of vital signs, environmental conditions, and physical strain [45]. The information gathered from this real-time data logging can be further processed into improvements of safety, ergonomics, and the overall well-being of the employees. One such wearable movement tracking sensor based on gallium thin films is displayed in Fig. 4.10.

(c) **Photonics in Quantum Sensing**: Another investigative area is the utilization of quantum photonics in achieving better sensitivity in chemical and physical sensors through entanglement [46]. On the other hand, by using quantum

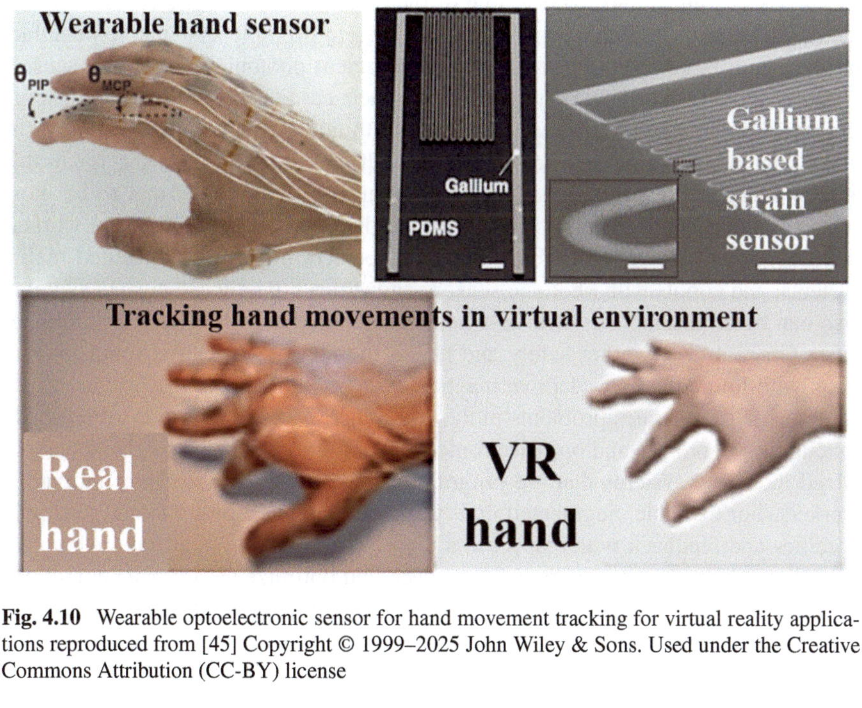


**Fig. 4.10** Wearable optoelectronic sensor for hand movement tracking for virtual reality applications reproduced from [45] Copyright © 1999–2025 John Wiley & Sons. Used under the Creative Commons Attribution (CC-BY) license

photonic sensors in a technically more precise way, one could very well pupate the age limits for measurement. At the same time, they are also capable of detecting quantum-level changes of physical parameters; thus, they have enormous significance for the territories as semiconductor manufacturing and advanced material science where the quantum-level precision is demanded.

(d) **IoT-Enabled Optoelectronic Sensors**: The integration of optoelectronic sensors along with the Internet of Things (IoT) is another major research area [47]. Smart optoelectronic sensors which are powered by IoT can pass and receive data in real time with other devices, which permits for intelligent, networked industrial systems. This opens up possibilities for even more automation, data-driven decision-making, and flexible responses to the changing industrial conditions. Research aims to come up with low-power and high-efficiency IoT-integrated optoelectronic sensors that are applicable in distant or hard-to-reach industrial locations.

### *4.5.3 Future Opportunities for Sustainable, Human-Friendly Industrial Environments*

The future prospects for photonics and optoelectronics in Industry 5.0 are immense. Technologies that will bring about this will be the ones that not only make industries more productive but also will create a living environment that takes into account the values of humankind. Indeed, one of the most promising opportunities is with the application of photonics and optoelectronics to promote sustainable manufacturing practices [48]. By embedding power-efficient photonic devices and sensors in the production line, industries can lessen their carbon footprint, reduce energy consumption, and minimize waste. Written with energy sources technologies, the technologies will be a key factor to help with such management the renewable energy sources like solar power which in return will help the planet to be more sustainable. Another clearest of all is the production of user-centric and worker-oriented industrial apparatus. By utilizing optoelectronic wear, augmented reality systems, and cobots with photonic sensors, industries not only can create safety but also can better ergonomics with less human intervention [49]. These technologies will offer real-time worker safety and health monitoring, industrial robot interactions with humans in an adaptive manner, and the implementation of data-driven solutions to ergonomics problems in the workplace. Moving forward, sophistically functioning photonics and optoelectronics technologies will also spearhead personalized industrial systems that are programmed to adapt to the needs of individual workers. For example, augmented reality systems could be programmed to supply specifics and feedback with respect to a worker's skill or physical state. Along the same line, photonic sensors are used to check and optimize the working conditions for a particular employee, thus creating a working environment that is individual-specific and more supported. The breakthrough in the integration of photonic and optoelectronic technologies with artificial intelligence (AI) will pave the way for intelligent industrial systems. AI algorithms can manipulate photon sensors and hence can process real-time data to induce more productivity, better predictions, and even autonomous decision-making. The AI-photonics combination will make it possible to create transformative self-adapted and self-controlled industrial environments, where robots and machines can continuously acquire more knowledge and adapt themselves to the upturning job market.

## 4.6 Conclusion

The integration of photonics and optoelectronics in Industry 5.0 intonates a revolutionary twist in the working operation of industries, which in turn offers great potential for increased innovation, efficiency, and sustainability. This chapter has discussed how these technologies, driven by advances in the control of light and the fusion of optical and electronic systems, are making way for a new industrial era which is not only technologically advanced but also human-centric and environmentally responsible. Photonics, as a technology that manipulates light for transmitting and processing of materials, is significantly changing smart manufacturing by increasing accuracy, efficiency, and flexibility in the production line. Optoelectronics, which is a combination of the light gear and electrical instruments, is also used as an instrument of connection and conversation that is leading to a faster data transmission and minimizing loss of energy. The integration of all these elements into Industry 5.0 results in the production of smart and flexible machines that have the ability to respond to real-time data, run themselves efficiently, and, thus, provide a more robust industrial setting. However, there are some challenges that needs to be dealt with in order to realize the full potential of photonics and optoelectronics. The expenses of execution, integration intricacy, and sustaining in industrial environments represent significant obstacles. Apart from that, the maintenance of data security and privacy among the ever-increasingly interconnected industrial systems is still a pressing issue. To overcome these technical challenges, continuous innovation and collaboration between industries, and also the development of photonic materials, sensor technologies, and AI integration, will be necessary. In general, photonics and optoelectronics in Industry 5.0 have a bright and promising future. Ongoing research of the new generation optoelectronic sensors, flexible photonic materials, and quantum sensing technologies present the further development of capabilities of these technologies. Further, the combination of photonics and AI and the Internet of Things will pave the way for the development of more intelligent and adaptive industrial systems that besides efficiency also give priority to the safety, health, and sustainability of the workers. In the area of sustainability, photonics and optoelectronics make a remarkable contribution to the environment by cutting down on the pollution caused by industrial operations. Energy-efficient manufacturing practices, optimized resource utilization, and enhanced management of renewable energy sources could be major players by providing greener industrial applications aligned with global sustainability objectives. In conclusion, the adoption of photonics and optoelectronics in Industry 5.0 will not only improve technological operations but will also ensure a sustainable, resilient, and fair environment in industries. Nevertheless, the integration of photonics, optoelectronics, AI, and IoT will bring about the advent of a new age in smart and human-friendly industrial landscapes. As research and development in these fields continue to progress, the next decade will most probably witness significant breakthroughs, further integrating these technologies into the fabric of industrial processes, making Industry 5.0 a reality.

## References

1. Leng J, Sha W, Wang B, Zheng P, Zhuang C, Liu Q, et al. Industry 5.0: prospect and retrospect. J Manuf Syst. 2022;65:279–95.
2. Ivanov D. The Industry 5.0 framework: viability-based integration of the resilience, sustainability, and human-centricity perspectives. Int J Prod Res. 2023;61(5):1683–95.
3. Demir KA, Döven G, Sezen B. Industry 5.0 and human-robot co-working. Procedia Comput Sci. 2019;158:688–95.
4. Luthra S, Mangla SK. Evaluating challenges to Industry 4.0 initiatives for supply chain sustainability in emerging economies. Process Saf Environ Protect. 2018;117:168–79.
5. Mourtzis D, Angelopoulos J, Panopoulos N. A literature review of the challenges and opportunities of the transition from Industry 4.0 to Society 5.0. Energies (Basel). 2022;15(17).
6. Xu X, Lu Y, Vogel-Heuser B, Wang L. Industry 4.0 and Industry 5.0—inception, conception and perception. J Manuf Syst. 2021;61:530–5.
7. Wosinski L, Thylén L. Integrated photonics in the 21st century. Photonics Res. [Internet]. 2014 Apr 1 [cited 2024 Sep 26];2(2):75–81. Available from: https://opg.optica.org/viewmedia.cfm?uri=prj-2-2-75&seq=0&html=true.
8. Deepak JR, Anirudh RP, Saran Sundar S. Applications of lasers in industries and laser welding: a review. Mater Today Proc. 2023.
9. Kaushik P, Singh DP, Rajpoot S. Fibre optic communication in 21st century. In: Proceedings of international conference on intelligent engineering and management, ICIEM 2020. 2020 Jun 1. pp. 125–9.
10. Sweeney SJ, Mukherjee J. Optoelectronic devices and materials. In: Springer handbooks [Internet]. 2017 [cited 2024 Sep 26];1–1. Available from: https://doi.org/10.1007/978-3-319-48933-9_35.
11. Massaro A. Advanced electronic and optoelectronic sensors, applications, modelling and Industry 5.0 perspectives. Appl Sci. 2023 [Internet]. 2023 Apr 4 [cited 2024 Sep 26];13(7):4582. Available from: https://www.mdpi.com/2076-3417/13/7/4582/htm.
12. Amiri IS, Azzuhri SR Bin, Jalil MA, Hairi HM, Ali J, Bunruangses M, et al. Introduction to photonics: principles and the most recent applications of microstructures. Micromachines. 2018 [Internet]. 2018 Sep 11 [cited 2024 Sep 26];9(9):452. Available from: https://www.mdpi.com/2072-666X/9/9/452/htm.
13. Jin L, Wang H, Cao R, Khan K, Tareen AK, Wageh S, et al. The rise of 2D materials/ferroelectrics for next generation photonics and optoelectronics devices. APL Mater [Internet]. 2022 Jun 1 [cited 2024 Sep 27];10(6):60903. Available from: https://pubs.aip.org/aip/apm/article/10/6/060903/2834982/The-rise-of-2D-materials-ferroelectrics-for-next.
14. Lu Y, Xu X, Wang L. Smart manufacturing process and system automation—a critical review of the standards and envisioned scenarios. J Manuf Syst. 2020;1(56):312–25.
15. Genna S, Menna E, Rubino G, Tagliaferri V. Experimental investigation of industrial laser cutting: the effect of the material selection and the process parameters on the Kerf quality. Appl Sci. 2020 [Internet]. 2020 Jul 19 [cited 2024 Sep 27];10(14):4956. Available from: https://www.mdpi.com/2076-3417/10/14/4956/htm.

16. Wong K V., Hernandez A. A review of additive manufacturing. Int Sch Res Notices [Internet]. 2012 Jan 1 [cited 2024 Sep 26];2012(1):208760. Available from: https://doi.org/10.5402/2012/208760.
17. Maharjan N, Wu N, Zhou W. Hardening efficiency and microstructural changes during laser surface hardening of 50CrMo4 steel. Metals. 2021 [Internet]. 2021 Dec 13 [cited 2024 Sep 27];11(12):2015. Available from: https://www.mdpi.com/2075-4701/11/12/2015/htm.
18. Dong T, Simões J, Yang Z. Flexible photodetector based on 2D materials: processing, architectures, and applications. Adv Mater Interfaces [Internet]. 2020 Feb 1 [cited 2024 Sep 26];7(4):1901657. Available from: https://doi.org/10.1002/admi.201901657.
19. Gupta M, Khan MA, Butola R, Singari RM. Advances in applications of non-destructive testing (NDT): a review. Adv Mater Process Technol [Internet]. 2022 Apr 3 [cited 2024 Sep 26];8(2):2286–307. Available from: https://doi.org/10.1080/2374068X.2021.1909332.
20. Brackett CA. Dense wavelength division multiplexing networks: principles and applications. IEEE J Sel Areas Commun. 1990;8(6):948–64.
21. Nagarajan R, Joyner CH, Schneider RP, Bostak JS, Butrie T, Dentai AG, et al. Large-scale photonic integrated circuits. IEEE J Sel Top Quantum Electron. 2005;11(1):50–62.
22. Cao H, Chriki R, Bittner S, Friesem AA, Davidson N. Complex lasers with controllable coherence. Nat Rev Phys. 2018 [Internet]. 2019 Jan 17 [cited 2024 Sep 26];1(2):156–68. Available from: https://www.nature.com/articles/s42254-018-0010-6.
23. Forrest SR. Optical detectors: three contenders. IEEE Spectr. 1986;23(5):76–84.
24. Wang C, Zhang M, Chen X, Bertrand M, Shams-Ansari A, Chandrasekhar S, et al. Integrated lithium niobate electro-optic modulators operating at CMOS-compatible voltages. Nature [Internet]. 2018 Sep 24 [cited 2024 Sep 26];562(7725):101–4. Available from: https://www.nature.com/articles/s41586-018-0551-y.
25. Cao R, Fan S, Yin P, Ma C, Zeng Y, Wang H, et al. Mid-infrared optoelectronic devices based on two-dimensional materials beyond graphene: status and trends. Nanomaterials [Internet]. 2022 Jul 1 [cited 2024 Sep 27];12(13):2260. Available from: https://www.mdpi.com/2079-4991/12/13/2260/htm.
26. Matera F. Photonics for the 5G era. Fiber Integr Optics [Internet]. 2017 Mar 4 [cited 2024 Sep 26];36(1–2):1–2. Available from: https://doi.org/10.1080/01468030.2017.1288980.
27. Chan VWS. Free-space optical communications. J Lightwave Technol. 2006;24(12):4750–62.
28. Moon SR, Cho SH, Sung M, Rha HY, Park KH, Lee JK, et al. 6G Indoor network enabled by photonics- and electronics-based sub-THz technology. J Lightwave Technol. [Internet]. 2022 Jan 15 [cited 2024 Sep 26];40(2):499–510. Available from: https://opg.optica.org/abstract.cfm?uri=jlt-40-2-499.
29. Akyildiz IF, Kak A, Nie S. 6G and beyond: the future of wireless communications systems. IEEE Access. 2020;8:133995–4030.
30. Shi Q, Dong B, He T, Sun Z, Zhu J, Zhang Z, et al. Progress in wearable electronics/photonics—moving toward the era of artificial intelligence and internet of things. InfoMat [Internet]. 2020 Nov 1 [cited 2024 Sep 26];2(6):1131–62. Available from: https://doi.org/10.1002/inf2.12122.
31. Laghari AA, Wu K, Laghari RA, Ali M, Khan AA. A review and state of art of Internet of Things (IoT). Arch Comput Meth Eng. [Internet]. 2022 May 1 [cited 2024 Sep 26];29(3):1395–413. https://doi.org/10.1007/s11831-021-09622-6. .
32. Aleksic S. A survey on optical technologies for IoT, smart industry, and smart infrastructures. J Sens Actuator Netw. [Internet]. 2019 Sep 17 [cited 2024 Sep 26];8(3):47. Available from: https://www.mdpi.com/2224-2708/8/3/47/htm.
33. Butt MA, Mateos X, Piramidowicz R. Photonics sensors: a perspective on current advancements, emerging challenges, and potential solutions (Invited). Phys Lett A. 2024;15(516):129633.
34. Alagappan G, Ong JR, Yang Z, Ang TYL, Zhao W, Jiang Y, et al. Leveraging AI in photonics and beyond. Photonics 2022;9:75 [Internet]. 2022 Jan 28 [cited 2024 Sep 26];9(2):75. Available from: https://www.mdpi.com/2304-6732/9/2/75/htm.
35. Shastri BJ, Tait AN, Ferreira de Lima T, Pernice WHP, Bhaskaran H, Wright CD, et al. Photonics for artificial intelligence and neuromorphic computing. Nature Photonics 2021 15:2 [Internet].

2021 Jan 29 [cited 2024 Sep 26];15(2):102–14. Available from: https://www.nature.com/articles/s41566-020-00754-y.
36. Kudyshev ZA, Shalaev VM, Boltasseva A. Machine learning for integrated quantum photonics. ACS Photonics [Internet]. 2021 Jan 20 [cited 2024 Sep 26];8(1):34–46. Available from: https://doi.org/10.1021/acsphotonics.0c00960.
37. Goda K, Jalali B, Lei C, Situ G, Westbrook P. AI boosts photonics and vice versa. APL Photonics [Internet]. 2020 Jul 1 [cited 2024 Sep 26];5(7). Available from: https://pubs.aip.org/aip/app/article/5/7/070401/1024610/AI-boosts-photonics-and-vice-versa.
38. Saito S. Editorial: integrated quantum photonics. Front Phys [Internet]. 2021 Nov 26 [cited 2024 Sep 26];9:800648. Available from: www.frontiersin.org.
39. Alahi MEE, Sukkuea A, Tina FW, Nag A, Kurdthongmee W, Suwannarat K, et al. Integration of IoT-enabled technologies and Artificial Intelligence (AI) for smart city scenario: recent advancements and future trends. Sensors 2023;23:5206 [Internet]. 2023 May 30 [cited 2024 Sep 27];23(11):5206. Available from: https://www.mdpi.com/1424-8220/23/11/5206/htm.
40. Moody G, Sorger VJ, Blumenthal DJ, Juodawlkis PW, Loh W, Sorace-Agaskar C, et al. Roadmap on integrated quantum photonics. J Phys: Photonics [Internet]. 2022 Jan 31 [cited 2024 Sep 26];4(1):012501. Available from: https://doi.org/10.1088/2515-7647/ac1ef4.
41. Mahmoud MG, Hares AS, Hameed MFO, El-Azab MS, Obayya SSA. AI-driven photonics: unleashing the power of AI to disrupt the future of photonics. APL Photonics [Internet]. 2024 Aug 1 [cited 2024 Sep 26];9(8):80902. Available from: https://pubs.aip.org/aip/app/article/9/8/080902/3309079/AI-driven-photonics-Unleashing-the-power-of-AI-to.
42. Soref R. The achievements and challenges of silicon photonics. Adv Opt Technol [Internet]. 2008 [cited 2024 Sep 26];7. Available from: www.mphotonics.mit.edu/about.
43. Kharche S, Dere P. Interoperability issues and challenges in 6G networks. J Mobile Multimedia [Internet]. 2022 [cited 2024 Sep 26];18(5):1445–70. Available from: https://www.researchgate.net/publication/359739223.
44. Chen W, Liu G. Intelligent optoelectronic devices for next-generation artificial machine vision. Adv Electron Mater [Internet]. 2022 Dec 1 [cited 2024 Sep 26];8(12):2200668. Available from: https://doi.org/10.1002/aelm.202200668.
45. Dejace L, Laubeuf N, Furfaro I, Lacour SP, Dejace L, Laubeuf N, et al. Gallium-based thin films for wearable human motion sensors. Adv Intell Syst [Internet]. 2019 Sep 1 [cited 2024 Sep 27];1(5):1900079. Available from: https://doi.org/10.1002/aisy.201900079.
46. Pirandola S, Bardhan BR, Gehring T, Weedbrook C, Lloyd S. Advances in photonic quantum sensing. Nature Photon [Internet]. 2018 Nov 28 [cited 2024 Sep 26];12(12):724–33. Available from: https://www.nature.com/articles/s41566-018-0301-6.
47. Singh KR, Nayak V, Singh J, Singh RP. Nano-enabled wearable sensors for the Internet of Things (IoT). Mater Lett. 2021;1(304): 130614.
48. Ghobakhloo M, Iranmanesh M, Foroughi B, Babaee Tirkolaee E, Asadi S, Amran A. Industry 5.0 implications for inclusive sustainable manufacturing: An evidence-knowledge-based strategic roadmap. J Clean Prod. 2023;417:138023.
49. Pizoń J, Cioch M, Kanski L, García ES. Cobots implementation in the era of Industry 5.0 using modern business and management solutions. Adv Sci Technol Res J. 2022;16(6):166–78.

# Chapter 5
# Implementing Photonics in Advanced Manufacturing Techniques for Industry 5.0

**Jyoti Katyal**

**Abstract** Photonics technology is key in numerous industrial operations which have merged to foster quicker, more accurate, and efficient manufacturing processes across various sectors. Photonics involves the manipulation, generation, and detection of both light and its properties which can result in significant improvements when leveraged in manufacturing. The paper explores how photonics is used in advanced manufacturing techniques, highlighting its important role in building Industry 5.0. High-speed optical communication networks, photonic sensors, and advanced laser technologies enable manufacturers to create interconnected and flexible production environments. This integration drives innovation, stimulates human–machine collaboration, and supports sustainability initiatives, all of which help accelerate the transition to a more intelligent and adaptable manufacturing environment that aligns with the goals of Industry 5.0.

**Keywords** Photonic · Industry 5.0 · Manufacturing · Sensing · Optical fiber

## 5.1 Introduction

Industry 5.0 is a paradigm shift in manufacturing, supply chain, and inventory paradigm that focuses on human-centered design. Industry 5.0 is an extension of the concepts established by Industry 4.0 with an emphasis on environmentally friendly, human-centric, and resilient systems. Among the key features of Industry 5.0 is its human-centric features focused on the ability to collaborate between humans and machines [1, 2]. This synergy boosts productivity and efficiency as well as creates new job opportunities. When you introduce digital twins and collaborative robots (cobots) into manufacturing workflows, it gives workers more flexibility and control over their roles. A healthy and happy workforce with opportunities for personal and professional development generates sustainable value for the organization. Resilience

J. Katyal (✉)
Amity Institute of Applied Science, Amity University, Noida, India
e-mail: jyoti_106@yahoo.co.in

 
V. Bhardwaj et al. (eds.), *Photonics and Optoelectronics in Industry 5.0*, Advances in Optics and Optoelectronics, https://doi.org/10.1007/978-981-96-7594-4_5

is also one of the other basic elements, which states that since Industry 5.0 must build resilience in the face of uncertainty [3, 4]. It calls for a proactive approach to strategy to anticipate potential disruptions across the value chain—inside the factory, out in supplier networks, and through transport networks. Digital technologies, simulations, and AI-driven modeling can enable companies to find the best possible alternatives in any disruption scenario based on cost, quality, logistics, etc. Businesses must identify their weakest links, set aside critical inputs, and devise contingency plans to respond to challenges effectively and maintain operational continuity. Last but not the least, Industry 5.0 also enhances sustainability within the manufacturing environment. Businesses are forced to assess their resource footprint due to growing regulatory scrutiny and investor demand to minimize environmental effect and cut carbon emissions. The sources of raw materials, waste production, environmental impacts, process energy efficiency, and energy source sustainability are all examined in this assessment. By following these guidelines, Industry 5.0 seeks to promote a manufacturing environment that places a high value on environmental sustainability, operational resilience, and human well-being [5]. Modern industrial processes rely heavily on photonics technology, which is revolutionizing a number of industries by increasing production speed, accuracy, and efficiency [6–9]. When incorporated into manufacturing processes, this field—which deals with the generation, manipulation, and detection of light and its properties—leads to significant breakthroughs. By incorporating photonics into advanced manufacturing for Industry 5.0, production processes can be revolutionized, leading to the development of smarter, more efficient, and sustainable manufacturing systems that leverage cutting-edge technologies [10, 11].

## 5.2 Key Aspects of Photonics in Advanced Manufacturing for Industry 5.0

In modern manufacturing, photonics is essential, particularly in the context of Industry 5.0, which prioritizes highly customized, sustainable, and human-centric production processes. High-precision manufacturing methods that minimize material waste and improve product quality are made possible by photonic technologies, such as laser machining and lithography. This is especially important in industries like aerospace, medical devices, and semiconductor manufacturing. With its remarkable accuracy for cutting, welding, and engraving, photonic lasers are used in material processing. These noncontact techniques reduce material waste, enhance product quality, and lessen tool wear [12, 13], Fig. 5.1. In sectors like aerospace, where even the smallest flaw could compromise functioning and safety, this degree of accuracy is essential. Photonics is a key component of extreme ultraviolet (EUV) lithography in the semiconductor industry, which is necessary to create semiconductors that are quicker, smaller, and use less energy. Photonic lithography guarantees accurate patterning on a nanoscale scale, which is crucial given the increasing need for electronics to be smaller. Photonics is used in health care to make complex parts of medical equipment while preserving sterility and accuracy during production. For the fabrication of complexly geometrized instruments and implants that are also biocompatible, laser machining is essential [14]. Photonics directly supports Industry 5.0 objectives and promotes overall sustainability by lowering material waste, improving product quality, and lowering the need for rework. In manufacturing settings, photonics-based sensors offer real-time feedback, facilitating predictive maintenance and more flexible production processes. By keeping an eye on a number of variables, including temperature, stress, and other important elements, these sensors can greatly increase manufacturing efficiency. By providing high-precision, real-time data that facilitate adaptive control and predictive maintenance, they play a crucial role in the development of intelligent, self-regulating production systems. This proactive strategy improves production procedures and helps avoid equipment failures. Predictive maintenance, where photonic sensors, such as fiber optic and laser-based systems, track vital parameters like vibration, temperature, pressure, and stress on machinery, is one of the key areas of effect. By detecting changes early, manufacturers can carry out maintenance before failures occur, reducing both downtime and repair costs. In terms of quality control, optical sensors integrated into production lines facilitate real-time assessments. For example, laser triangulation sensors can detect surface defects, ensuring that products meet exact specifications without slowing down production speed. Additionally, photonic sensors play a vital role in environmental monitoring by tracking parameters like gas emissions, humidity, and temperature within the factory. These data help optimize energy usage and support sustainability initiatives by minimizing waste and reducing energy consumption. While identifying changes promptly, manufacturers can perform maintenance before failures happen, decreasing both downtime and repair expenses. Regarding quality control, optical sensors incorporated into production lines enable immediate evaluations. For instance, laser triangulation sensors are capable of identifying surface imperfections, guaranteeing that items adhere to precise standards without hindering production pace. Moreover, photonic sensors are essential for environmental monitoring by measuring factors such as gas emissions, humidity, and temperature inside the factory. This information aids in optimizing energy consumption and promotes sustainability efforts by decreasing waste and lowering energy use. Additionally, by utilizing sensor data in AI and machine learning algorithms, production lines can autonomously modify operational parameters—like speed, temperature, or pressure—to improve efficiency and ensure uniform product quality, even during varying conditions. In conclusion, smart sensing utilizing photonics promotes a dynamic, data-oriented manufacturing environment that improves the collaboration between humans and machines essential to Industry 5.0 [15–17]. Photonics, particularly with laser-driven 3D printing, is transforming additive manufacturing by allowing the creation of complex shapes and tailored parts. This method aligns seamlessly with the innovation and adaptability needed in Industry 5.0. Methods such as selective laser sintering (SLS) and selective laser melting (SLM) employ powerful lasers to bond powdered substances layer by layer, facilitating the production of intricate shapes and structures that conventional subtractive techniques cannot produce, thereby broadening design opportunities. Moreover, the accuracy of laser-based additive manufacturing enables the production of custom components designed to meet specific customer needs, which is especially advantageous in industries such as aerospace, automotive, and medical, where customized parts are crucial for performance and proper fit [18, 19]. Moreover, photonics supports a diverse selection of materials, such as metals, polymers, and ceramics, providing manufacturers the opportunity to investigate novel applications and create components with distinctive traits like lighter designs and improved durability. Utilizing photonics in additive manufacturing greatly shortens the duration required to transition from design to production, fostering rapid prototyping that speeds up innovation cycles and allows businesses to swiftly adjust to market fluctuations. Additionally, additive manufacturing usually produces less waste compared to conventional manufacturing methods, since it utilizes only the required materials for every part and enables the recycling of surplus powders, thereby fostering sustainability. By utilizing the benefits of photonics in additive manufacturing, sectors can improve their creativity, productivity, and sustainability while completely adopting the human-focused and flexible concepts of Industry 5.0 [20]. Incorporating photonics into manufacturing methods is essential for improving energy efficiency and advancing sustainability. A key benefit of photonic technologies is their capacity to lower energy usage. Methods like laser cutting, welding, and drilling generally use less energy than conventional mechanical techniques, since lasers concentrate energy directly on the material, promoting quicker and more efficient processing that eventually reduces total energy expenses. Moreover, photonic techniques produce less heat than traditional methods, which frequently result in considerable thermal deformation and the necessity for elaborate cooling systems. For example, laser welding creates focused heating, which limits heat-affected areas and decreases the energy needed for cooling. Furthermore, photonics facilitates accurate material processing, allowing for the utilization of lighter, more sustainable materials without losing strength. This is especially crucial in sectors such as automotive and aerospace, where lighter parts can enhance fuel efficiency and lower emissions. Techniques in photonic manufacturing aid in minimizing waste by creating components that generate little scrap; for instance, laser cutting yields cleaner edges, aligning with sustainability concepts centered on conserving resources. Additionally, utilizing energy-saving photonic methods allows producers to reduce their carbon emissions and aids in wider sustainability efforts. This method aligns with the human-centered principles of Industry 5.0, which promotes responsible innovation and care for the environment. Moreover, photonics enables the integration of renewable energy via solar cells and various photonic devices, permitting manufacturers to include sustainable energy sources in their processes. In summary, utilizing photonics in production improves process efficiency and supports the sustainability objectives of Industry 5.0, promoting a more eco-friendly and responsible manufacturing setting [21–24]. Photonics technologies greatly improve human–machine interaction, promoting better cooperation between people and automated systems. These technologies facilitate a human-focused strategy in Industry 5.0 in various manners. Initially, photonics enables rapid optical communication, allowing devices to transfer data swiftly and securely.

This instantaneous data transmission enhances the responsiveness of manufacturing systems, granting workers quick access to essential information for prompt decision-making. Furthermore, holographic and augmented reality (AR) screens driven by photonics develop engaging visual interfaces for users. These screens cast digital data onto actual surroundings, aiding employees in visualizing intricate information, assembly guidelines, and upkeep activities, thus enhancing their understanding and productivity. Additionally, sophisticated user interfaces enabled by photonics facilitate the creation of intuitive systems that adjust to user preferences. For instance, utilizing optical sensors for touchless gesture control allows workers to engage with machines without any physical touch, improving hygiene and reducing wear on equipment. Photonics improves training and simulation abilities by enabling manufacturers to develop lifelike training settings via virtual and augmented reality. This method allows employees to hone their abilities in a secure environment, connecting theoretical understanding with hands-on experience, which leads to a more proficient workforce. Additionally, photonics boosts the safety and efficiency of collaborative robots (cobots) by improving their sensing abilities. Optical sensors enable these robots to identify the presence and closeness of human operators, promoting safer interactions in collaborative work environments. Ultimately, incorporating photonics into human–machine interfaces enables workers to harness their creativity and intuition in conjunction with automated systems. This partnership fosters creativity, resulting in original answers to intricate manufacturing difficulties. In conclusion, photonics-based human–machine interaction fosters a cooperative setting where the strengths of technology and human insight are utilized efficiently. This aligns seamlessly with the objectives of Industry 5.0, fostering a more creative, efficient, and human-centric manufacturing environment [25–28].

## 5.3 Challenges of Implementing Photonics in Advanced Manufacturing

The integration of photonics technologies in production faces various obstacles linked to material characteristics and technical intricacies [29]. Various materials have distinct optical properties, including absorption, reflection, and refraction, which can make the use of certain photonic methods like laser cutting or welding more complex. Moreover, certain materials are sensitive to heat and might not endure processes that produce considerable thermal energy, requiring precise adjustment of photonic methods to avoid harm while attaining the intended outcomes. Surface properties, such as texture and finish, can influence the efficiency of photonic processes; impurities or uneven surfaces may interfere with laser focus, impacting precision and quality. In the medical device industry, materials are required to adhere to strict biocompatibility standards, which makes the incorporation of photonics more challenging to guarantee that no harmful residues are left after processing. From a technical perspective, incorporating photonic technologies into current manufacturing systems can be

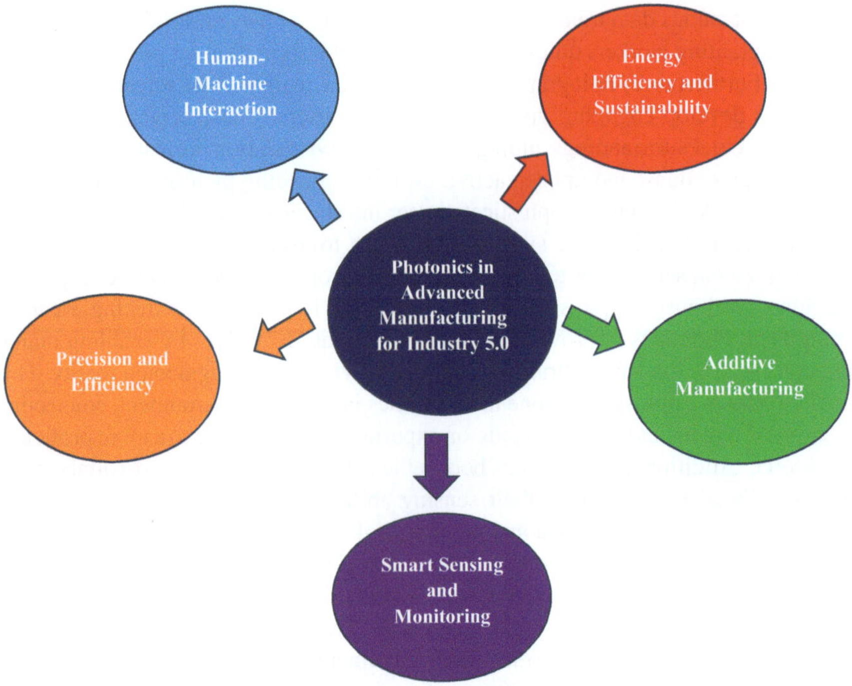


**Fig. 5.1** Key aspects of photonics in advanced manufacturing for Industry 5.0

a challenging task. Producers might have to spend on new machinery and training, potentially causing interruptions in operations during the shift. Enhancing photonic processes demands significant experimentation and adjustments to attain optimal settings for power, speed, and focus, which can be both time-consuming and resource-heavy. The upfront expenses linked to advanced photonics tools, like laser systems and optical sensors, can pose a challenge, especially for smaller manufacturers who might find these investments difficult. Continuous upkeep and servicing of photonic systems may lead to extra costs, necessitating skilled technicians and spare components. Additionally, there is frequently a gap in skills among workers concerning knowledge in photonics applications in manufacturing. Training current employees or bringing in new talent can require significant resources. Ongoing research and development are essential for staying current with advancements in photonics technology, requiring additional investment to foster innovation and remain competitive. Adhering to regulatory standards poses another challenge, especially in sectors such as medical devices and aerospace, where stringent regulations must be satisfied. This frequently entails thorough documentation and testing, and the certification procedures for emerging photonic technologies can be time-consuming and intricate, possibly postponing their adoption in production. Finally, producers need to take into account the ecological effects of photonic processes, especially regarding energy usage and waste production. It is crucial that photonics technologies are in harmony with sustainability objectives. By tackling these issues, manufacturers can successfully utilize photonics to transform their production methods, improve efficiency, and promote innovation in advanced manufacturing. Incorporating photonics technologies into current manufacturing processes and machinery can present considerable difficulties, since numerous existing systems are not built to support these sophisticated tools [30]. This frequently requires significant changes or even full substitutions of machinery. Additionally, manufacturing usually includes multiple processes like cutting, welding, and assembly, making it challenging to align photonics-based methods with conventional techniques for smooth production. Meticulous planning and oversight are crucial for attaining this integration. Furthermore, the integration of photonics technologies necessitates a reassessment of current workflows. Producers need to plan ways to integrate photonic processes successfully to improve efficiency and reduce downtime during shifts. Data integration presents another challenge, since photonics systems often produce large volumes of data. Integrating this information with existing management frameworks—such as Enterprise Resource Planning (ERP) or Manufacturing Execution Systems (MES)—can present difficulties, especially if the current systems are incompatible or not designed for real-time data analysis. Finally, as advanced manufacturing becomes more dependent on automation and robotics, it is crucial to ensure that photonic technologies can interact smoothly with automated systems to maintain productivity and efficiency in the manufacturing setting. The effectiveness of translating optical signals into electrical signals, and the reverse, poses a major challenge in photonic systems. Inefficient conversion can lead to signal loss and increased energy usage, negatively impacting the overall performance of the system. Furthermore, ensuring signal integrity in this conversion process is vital, since any loss of information may jeopardize data accuracy and dependability in manufacturing activities. Integration of new photonic technologies with current electrical systems might lead to compatibility problems, possibly obliging manufacturers to invest in new interfaces or adapters, which could raise costs and complexity. Creating efficient optical-electrical interfaces requires specialized knowledge, posing a challenge for manufacturers without in-house photonics expertise. Thermal management is another key aspect, since optical–electrical interfaces may produce heat, particularly during rapid data transfer. Effective heat management is crucial for guaranteeing the durability and dependability of both photonic and electronic components. Additionally, merging photonics with electronic systems necessitates meticulous planning to guarantee seamless interaction between these systems, which can be especially challenging when aligning optical and electronic signals. The physical dimensions of optical–electrical interfaces can create difficulties in tightly confined manufacturing settings, making it essential to develop compact and effective solutions to reduce spatial needs. Ultimately, the lack of standardized optical–electrical interface technologies may result in interoperability challenges between various systems and components, emphasizing the necessity for industry standards to encourage the wider implementation of photonics in manufacturing. Translating photonics-based methods from laboratory settings to large-scale production poses considerable challenges. Methods that excel in small batch environments may struggle to transition to high-volume production, requiring careful reassessment and adjustments. Moreover, photonics generally encompasses intricate multistep procedures, including cutting, welding, and modifying surfaces. Overseeing these complex process chains raises the chances of mistakes, lengthens cycle times, and makes it more difficult to uphold uniform quality throughout every stage of production. Additionally, numerous manufacturing plants continue to rely on conventional techniques, and incorporating photonics into these established processes can create challenges, as the two methods might not fit together smoothly regarding speed, quality, and efficiency. Although photonics technologies excel in producing tailored products, this ability to customize can add additional complexity to the manufacturing process. Effectively managing the requirements for customization while ensuring efficient mass production necessitates meticulous planning and proper allocation of resources. Although photonics technologies can improve efficiency in specific uses, they frequently entail considerable energy needs, especially for running high-power lasers and other photonic systems. This notable energy demand can result in higher operational expenses, possibly negating some of the efficiency advantages provided by these technologies. Moreover, numerous photonic systems, particularly high-powered lasers, generate significant heat while functioning, requiring efficient cooling methods. These cooling systems generally use additional energy, which adds to total power consumption. Creating photonic systems that successfully optimize performance while maintaining energy efficiency poses a challenge for engineers, who need to thoroughly assess power needs throughout the design phase to prevent unnecessary energy consumption. This confrontation raises worries for numerous producers seeking to incorporate eco-friendly technologies. In extensive manufacturing settings, significant energy usage can raise operational expenses, potentially dissuading certain firms from embracing cutting-edge photonics technologies. Additionally, dependence on stable and trustworthy energy sources can pose challenges, especially in areas with unreliable power supplies, which may interrupt the seamless functioning of photonics-based systems. Ultimately, producers need to manage environmental rules concerning energy usage and effects. Excessive energy consumption could lead to violations of local laws or sustainability goals, necessitating additional investment in energy-saving alternatives.

## 5.4 Applications

Cameras and sensors utilizing photonics are crucial for quality assurance, evaluation, and measurement in manufacturing processes, Fig. 5.2. These technologies play a crucial role in various fields. For example, high-resolution photonic cameras with sophisticated imaging methods such as hyperspectral or multispectral imaging allow for accurate examination of items on assembly lines. They are capable of identifying surface flaws, color variations, and structural anomalies instantly, guaranteeing that only products that satisfy quality standards advance in the manufacturing process. Furthermore, photonic techniques, such as laser-induced fluorescence and optical coherence tomography, offer nondestructive testing alternatives that enable thorough examination of materials and parts without inflicting any damage [31, 32]. This ability is especially important in sectors like aerospace and automotive, where preserving part integrity is essential. Regarding precision measurement, photonics is crucial in metrology, employing methods such as laser triangulation and interferometry to provide high accuracy in assessing distances, angles, and surface profiles. This guarantees correct fitting and operation of components. In addition, photonics enables the creation of intricate 3D models using structured light and laser scanning techniques, which are essential for confirming geometric precision and aiding reverse engineering processes. The combination of photonics-driven imaging systems with machine learning and AI enables automated examination and assessment, improving efficiency and reducing dependence on manual inspections. Real-time imaging offers essential insights into manufacturing processes, allowing for instant modifications for optimization, like adjusting thermal profiles during laser cutting. Through the use of photonics-based imaging technologies, producers can greatly improve their quality assurance and inspection procedures, resulting in higher product quality, reduced waste, and enhanced overall efficiency. This ability advances the objectives of Industry 5.0 by encouraging innovation and upholding high standards in a competitive marketplace [33, 34]. Photonics plays a crucial role in sensing applications within manufacturing, enabling precise oversight and regulation of key parameters for ideal production processes. Photonic sensing technologies provide multiple significant benefits to manufacturing [35, 36]. For example, fiber optic temperature sensors deliver accurate and quick real-time temperature measurements, essential in procedures that need stringent thermal regulation, like semiconductor manufacturing and material handling. Moreover, fiber optic sensors are capable of measuring strain and stress in materials and structures. By integrating these sensors into structures, producers can constantly evaluate the condition of essential components, facilitating predictive maintenance to avert failures before they happen. Photonic sensors are also highly effective in detecting chemicals and gases, employing methods such as photoacoustic spectroscopy and surface plasmon resonance to track specific substances with great sensitivity, thus ensuring safety and adherence in settings that handle hazardous materials. Furthermore, laser-based sensors are extensively utilized for accurate position and displacement measurements, able to sense tiny movements, which makes them perfect for robotics and automation applications that demand precise control. Photonic sensors additionally track vibrations in equipment, enabling manufacturers to examine vibration patterns for early detection of possible mechanical problems, aiding in prompt maintenance and minimizing downtime. Furthermore, photonics allows precise assessment of liquid levels and flow rates via noninvasive optical sensors, ensuring ongoing surveillance in processes that deal with liquids or slurries. These sensing technologies can be effortlessly combined with sophisticated data analytics systems, enabling real-time data gathering and examination. This ability enables manufacturers to improve processes using real-time data, increasing efficiency and decision-making. Utilizing photonic sensing technologies enables manufacturers to enhance precision and dependability in tracking different parameters, leading to better process management, less waste, and heightened overall productivity. This is consistent with the concepts of Industry 5.0, which focuses on a data-oriented and flexible method for manufacturing [37, 38]. Photonics technologies are progressively employed in diverse material processing tasks in manufacturing, allowing for accurate and effective handling of materials [39]. A key use is laser cutting, enabling remarkable accuracy and rapidity in cutting various materials. This technique reduces kerf (the breadth of the cut) and creates smooth edges, making it suitable for various materials, such as metals, plastics, and composites. As a result, laser cutting enables complex patterns and minimizes waste. Another important application is laser welding, which utilizes sophisticated methods to produce robust, high-quality connections while reducing heat-affected areas. This is especially beneficial in sectors like automotive and aerospace, where it is crucial to uphold structural integrity. Moreover, laser welding allows for the connection of different materials, further broadening design options. Photonics is essential in additive manufacturing, utilizing techniques such as selective laser melting (SLM) and selective laser sintering (SLS). These methods employ lasers to meld powdered substances layer by layer, enabling the formation of intricate shapes that conventional manufacturing processes frequently fail to accomplish. In surface treatment, photonics technologies like laser ablation and laser surface modification are used to change the surface characteristics of materials. This can boost adhesion, enhance wear resistance, or alter surface texture, which is especially advantageous in sectors that need specialized coatings or finishes. Micromachining represents another field where lasers offer accurate processing, allowing manufacturers to produce detailed designs on tiny parts. This ability is vital in fields such as electronics, where miniaturization is necessary, and components need to be manufactured with great precision. Additionally, photonics is employed for precisely engraving and marking materials. Laser marking techniques can produce permanent markings on different surfaces such as metals, plastics, and glass, without changing the material properties, which makes them perfect for branding and traceability. Finally, laser thermal processing methods, including laser annealing and hardening, enable targeted heating and cooling of materials. This method can improve mechanical characteristics and efficiency while reducing distortions typically linked to traditional heat treatment. Utilizing photonics for material processing allows manufacturers to attain improved precision, efficiency, and adaptability in their production methods. This integration is in line with the tenets of Industry 5.0, fostering innovation and flexibility in a more competitive manufacturing environment [40–42]. Cutting-edge photolithography methods are essential in the semiconductor manufacturing process, facilitating the creation of next-generation microchips [43, 44]. Photonics, particularly via photolithography, plays an important role in this area in multiple ways. Initially, photolithography employs light to cast detailed circuit designs onto semiconductor wafers. Sophisticated photonic systems employ shorter wavelengths, like extreme ultraviolet (EUV) lithography, enabling greater resolution and reduced feature sizes in microchips, which is crucial for developing more compact and efficient components. The procedure starts with the coating of a light-sensitive substance, referred to as photoresist, onto the wafer. When this substance is subjected to light via a mask, the areas that are exposed experience chemical transformations that enable accurate pattern transfer—crucial for the intricate structure of contemporary microchips. Advancements in photonics have also resulted in the creation of sophisticated mask technologies that improve pattern accuracy and minimize defects throughout the lithography process. Methods such as phase-shifting masks and optical proximity correction enhance the precision of pattern transfer, guaranteeing that the intended features are faithfully reproduced on the wafer. Moreover, photonic technologies are utilized in advanced alignment systems to guarantee accurate stacking of several circuit layers on a wafer. High-accuracy optical sensors allow for precise alignment, essential for the overall performance and functionality of microchips. Optical metrology methods like reflectometry and ellipsometry are employed for monitoring and controlling processes [45, 46]. These techniques enable producers to oversee and regulate the thickness of the photoresist and additional layers during production, offering immediate feedback that guarantees uniform quality and reduces defects. Additionally, cutting-edge photolithography methods enhance the efficiency of semiconductor production while ensuring high yields, resulting in lower manufacturing costs and the capacity to fulfill the growing need for more powerful and efficient microchips. As the sector moves toward smaller process nodes (such as 5 nm and beyond), the importance of photonics in semiconductor manufacturing grows increasingly essential. Advancements in lithography are crucial for maintaining the trend of miniaturization and improving performance in electronic devices. In conclusion, photonics acts as a fundamental technology in the production of semiconductors, especially via sophisticated photolithography methods. These advancements are crucial for creating next-generation microchips that energize contemporary electronics, corresponding with Industry 5.0 objectives by promoting technological progress and improving manufacturing capabilities. These technologies play a crucial role in various important domains [47, 48]. To start, lasers offer precise cutting abilities, enabling manufacturers to produce detailed shapes and designs in different materials such as metals, polymers, and ceramics. This accuracy is crucial for items such as implants and prosthetics, where precise measurements and surface treatments are vital for performance and biocompatibility. One more benefit of laser cutting is its reduced thermal effect. In contrast to conventional mechanical cutting techniques that generate considerable heat, laser cutting minimizes thermal effects, which is especially crucial in the production of medical devices. This feature aids in maintaining the properties of heat-sensitive materials and protects adjacent areas from harm. Photonics enables the flexible handling of various materials utilized in medical devices. For example, lasers are capable of efficiently cutting and welding titanium and cobalt-chromium alloys typically used in orthopedic implants, along with biocompatible polymers utilized in prosthetic devices. Moreover, laser therapies can improve the surface characteristics of medical instruments. Methods like laser ablation enhance surface roughness, aiding in improved adhesion of coatings and increasing wear resistance in high-contact zones—essential factors for the enduring success of implants. Laser welding and bonding represent additional areas where photonics excels. These sophisticated methods form robust, exact joints without requiring extra adhesives or mechanical fasteners, essential for preserving the structural integrity of surgical instruments and implants. Additionally, the accuracy of lasers facilitates micromachining, enabling the creation of small medical devices and parts. This feature is particularly crucial for uses like catheters, stents, and minimally invasive surgical tools. Incorporating photonic systems into automated manufacturing lines improves both efficiency and consistency in the production of medical devices. This automation minimizes human mistakes, boosts production speeds, and guarantees top-notch results. Ultimately, photonics facilitates speedy prototyping and personalization of medical devices customized for specific patient requirements. For instance, 3D laser printing methods can produce tailored implants and prosthetics that exactly conform to a patient's anatomy. To summarize, lasers and photonics play a crucial role in precision manufacturing within the medical device industry. They support the creation of dependable, high-quality medical devices that enhance patient results while adhering to the tenets of Industry 5.0, which highlights innovation and flexibility [49]. Photonics plays a crucial role in creating high-speed optical fiber communication systems that enable immediate data transfer in smart factories [50, 51]. Here is how photonics improves communication systems in manufacturing: First, optical fibers provide much higher bandwidth than conventional copper cables, enabling fast transfer of vast data amounts. This trait renders them especially apt for the data-heavy tasks prevalent in smart factories. Additionally, photonic technologies facilitate long-distance data transmission with little signal loss, which is essential for interlinking various parts of a manufacturing site or connecting distant locations. This guarantees smooth communication across the whole network. Optical fiber communication systems additionally offer minimal latency in data transfer, an essential element for real-time uses in smart manufacturing. This rapid data transfer facilitates swift decision-making and improves the agility of automated systems. Moreover, photonic communication systems show enhanced reliability because they are less susceptible to electromagnetic interference and environmental influences compared to conventional electrical systems. This dependability guarantees steady and uniform communication, crucial for automated manufacturing systems and linked devices. Scalability represents another key benefit; optical fiber networks can be effortlessly extended to meet the increasing data needs of contemporary manufacturing plants. As production demands grow, manufacturers can enhance their communication systems without major renovations. Additionally, photonics aids in the incorporation of Internet of Things (IoT) devices within smart factories. High-speed optical communication allows IoT sensors and devices to transmit data effectively, which improves monitoring, control, and automation in production processes [29, 30, 52]. The elevated data rates enabled by optical fiber communication also aid advanced technologies such as big data analytics, artificial intelligence (AI), and machine learning in manufacturing, as these technologies necessitate strong data infrastructures for effective operation. Ultimately, optical fiber communication provides improved security attributes, making it difficult to tap into fiber optic cables undetected. This protection is essential for securing sensitive production information and preserving operational reliability. In conclusion, photonics is essential for developing optical fiber communication systems that facilitate fast, dependable, and scalable data transfer in smart factories. This ability corresponds with the tenets of

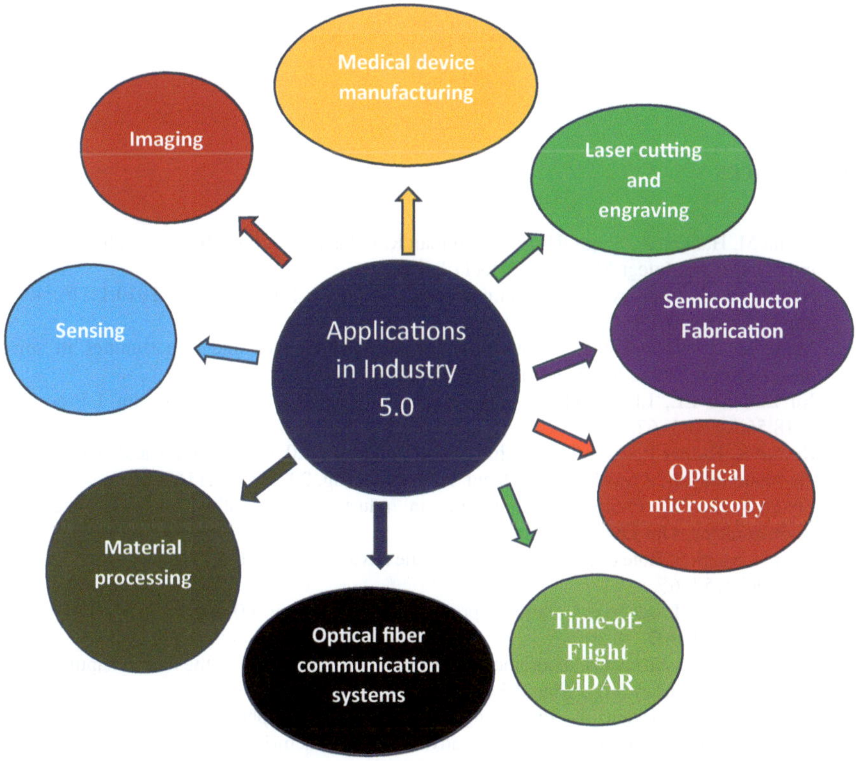


**Fig. 5.2** Applications of photonics in Industry 5.0

Industry 5.0, in which connectivity, immediate data evaluation, and automation are crucial for promoting innovation and improving manufacturing effectiveness.

## 5.5 Conclusion

The incorporation of photonics into cutting-edge manufacturing techniques signifies a revolutionary move toward Industry 5.0, focusing on sustainability, creativity, and human-centered strategies. By utilizing the potential of photonic technologies—like rapid optical communication, precise laser processing, and sophisticated sensing—producers can greatly improve efficiency, quality, and flexibility. Essential applications in multiple industries illustrate that photonics not only enables instantaneous data transfer and smooth communication but also enhances material processing and guarantees strict quality assurance. In addition, the collaboration between people and machines is enhanced by photonics-based interfaces and automation, promoting teamwork that emphasizes creativity and solving problems. As sectors keep adopting these innovations, the continuous progress and use of photonics will be crucial in forming a more efficient, sustainable, and creative future for manufacturing, ultimately aiding the broader goals of Industry 5.0.

## References


1. Javaid M, Haleem A, Singh RP, Rab S, Suman R. Industry 5.0: potential applications in smart industries. J Ind Integr Manag. 2021;6(1):1–15.
2. Nahavandi S. Industry 5.0—A human-centric solution. Sustainability. 2019;11(16):4371. https://doi.org/10.3390/su11164371.
3. Mourtzis D. Towards the vision of Industry 5.0: challenges and opportunities in smart manufacturing. Annu Rev Control. 2022;53:234–45.
4. Xu LD, Xu EL, Li L. Industry 4.0: state of the art and future trends. Int J Prod Res. 2018;56(8):2941–62.
5. Kumar R, Kumar P, Joshi G. The shift towards Industry 5.0: opportunities and challenges in the future industrial revolution. J Manuf Technol Manag. 2022;33(7):1169–85.
6. Zhang J, Guo Q. Photonics technologies in Industry 5.0: a pathway towards intelligent manufacturing. Opt Eng. 2020;59(4):045105.
7. Kaur H, Singh S. Role of photonic sensors in the advancement of Industry 5.0. J Sens Technol. 2021;9(3):157–65.
8. Shen Y, Li X. Laser-based photonic processing for Industry 5.0 applications. Photon Res. 2019;7(8):894–902.
9. Takahashi R, Nishizawa A. Photonic sensors and their integration in the smart manufacturing paradigm of Industry 5.0. J Manuf Syst. 2020;54:153–62.
10. Garzón B, Campuzano F. Enhancing Industry 5.0 with photonic technologies: High-speed communication and precision in advanced manufacturing. J Photonics Adv Technol. 2022;33(5):45–60.
11. Watanabe K, Mori T. Photonic solutions for the next-generation smart factories: Integration of photonics in Industry 5.0. Int J Optomech. 2021;11(3):198–210.
12. Meijer J, Du K, Gillner A, Hoffmann D, Kovalenko V, Masuzawa T, et al. Laser machining by short and ultrashort pulses, state of the art and new opportunities in the age of Industry 5.0. CIRP Ann. 2002;51(2):531–50.
13. Poprawe R. Industrial photonics in manufacturing: a key enabler for high-precision and efficient production. Proc IEEE. 2011;99(7):1206–13.
14. Nguyen DK, Madsen S, Petersen HM. Ultrafast laser processing for precision manufacturing in the era of Industry 5.0. J Manuf Process. 2020;58:619–30.
15. Rangin M, Manickam SK, Zhang Y. Monitoring for predictive maintenance. Opt Fiber Technol. 2018;45:217–27.
16. Grattan KTV, Meggitt BT. Optical fiber sensor technology: fundamentals and applications. Vol. 1. Springer; 2013.
17. Ma Y, Gao RX. Photonics-based smart sensing systems for monitoring in advanced manufacturing. IEEE Sens J. 2020;20(14):7528–37.
18. Gibson I, Rosen DW, Stucker B. Additive manufacturing technologies: 3D printing, rapid prototyping, and direct digital manufacturing. 2nd ed. Springer; 2014.
19. Yasa E, Kruth JP. Microstructural investigation of selective laser melting 316L stainless steel parts exposed to laser re-melting. Procedia Eng. 2011;19:389–95.
20. Thompson SM, Bian L, Shamsaei N, Yadollahi A. An overview of direct laser deposition for additive manufacturing; Part I: Transport phenomena, modeling, and diagnostics. Addit Manuf. 2015;8:36–62.
21. Duflou JR, Despeisse M, De Witte M. Manufacturing strategies in the context of sustainability: the case of photonics. In: Costa KAR, editor. Proceedings of the 19th CIRP international conference on life cycle engineering. 2012. pp. 178–83.

22. Cheng HK, Wong CH. Energy-efficient manufacturing and production: a review. Energy Rep. 2017;3:53–64.
23. Müller M, Schubert M. Photonic systems for energy-efficient manufacturing. Opt Express. 2015;23(8):10253–8. https://doi.org/10.1364/OE.23.010253.
24. Kopacek P, Škorpil J. Photonics for sustainable manufacturing and environmental monitoring. J Manuf Sci Eng. 2018;140(4):1–14.
25. Huang GQ, Hsu HK. Human-robot collaboration in intelligent manufacturing: a review. Robot Comput Integr Manuf. 2017;43:131–41.
26. Brewster S, Brown L. Integrating human-centered design in Industry 5.0 manufacturing systems. Hum-Centric Comput Inf Sci. 2019;9(1):1–16.
27. Lee J, Kwon OK. Augmented reality and virtual reality technologies in Industry 5.0: a review. J Manuf Syst. 2018;48:125–32.
28. Bader A, Haeusler K. Collaborative human-robot interaction in smart manufacturing environments: challenges and solutions. J Intell Manuf. 2020;31(6):1423–42.
29. Singh S, Verma S. Compact wearable dual wideband circularly polarized L-strip fed slotted textile antenna with parasitic elements for ISM band applications. J Electromagn Waves Appl. 2020;35(2):185–97.
30. Singh S, Verma S. Compact dual band and dual polarized CPW-fed triangular shaped wearable antenna for ISM band applications. Electromagnetics. 2022;42:38–50.
31. Kokkinos AI, Mavridis I. Real-time imaging in Industry 5.0: a framework for intelligent manufacturing. J Manuf Syst. 2020;54:162–75.
32. Bottani E, Rizzi A. The role of imaging technologies in Industry 4.0 and 5.0: a review of recent developments. Comput Ind. 2018;99:21–31.
33. Hahn J, Kim KJ. Enhancing manufacturing processes through advanced imaging technologies in Industry 5.0. Int J Adv Manuf Technol. 2021;115(1–4):235–45.
34. Yao Y, Wang H. Imaging and visualization technologies in Industry 5.0: challenges and opportunities. J Intell Manuf. 2022;33(4):1029–41.
35. Schneider J, Dörflinger D. Photonics in Industry 5.0: Revolutionizing smart sensors for manufacturing applications. J Photonics Energy. 2020;10(1):1–14.
36. Dai H, Zhang J. The role of photonic sensors in Industry 5.0: advances and future trends. Sensors. 2021;21(18):1–22.
37. Patel AR, Bhatia A. Smart photonic sensors for Industry 5.0: innovations and applications. Opt Express. 2022;30(12):21243–60.
38. Liu Y, Wang Y. Photonic sensor networks for intelligent manufacturing in Industry 5.0. IEEE Trans Ind Inform. 2023;19(3):1872–80.
39. Kumar V, Gupta A. Advancements in photonic material processing technologies for smart manufacturing in Industry 5.0. Mater Sci Eng A. 2021;814:141206.
40. He H, Wu Z. Laser processing of materials for Industry 5.0: challenges and opportunities. J Laser Appl. 2022;34(4):042001.
41. Pérez R, Garzón F. Photonics-driven material processing technologies for the future of smart manufacturing. Opt Lasers Eng. 2023;153: 107380.
42. Miao C, Zhao Q. The role of photonics in material processing for sustainable manufacturing in Industry 5.0. Sustain Mater Technol. 2020;25:e00175.
43. Lee J, Choi H. Advances in photonic semiconductor fabrication for next-generation electronics in Industry 5.0. J Semicond Technol Sci. 2021;21(3):269–80.
44. Hwang C, Park J. Photonic integration and semiconductor manufacturing: Pathways towards Industry 5.0. IEEE Trans Nanotechnol. 2022;21(5):896–903.
45. Khan A, Shah A. The impact of photonics on semiconductor fabrication techniques for smart manufacturing systems. Appl Phys Rev. 2023;10(1): 011318.
46. Zhang Y, Huang X. Emerging photonic technologies for semiconductor fabrication in the context of Industry 5.0. Mater Today Proc. 2020;33:234–40.
47. Zhao Y, Wang Q. Advancements in photonic technologies for the manufacturing of medical devices in Industry 5.0. J Manuf Process. 2021;64:198–210.

48. Singh R, Tiwari M. Integrating Industry 5.0 principles in the manufacturing of smart medical devices. Int J Med Robot Comput Assist Surg. 2022;18(2):e2333.
49. López J, Ortega A. The role of advanced manufacturing technologies in the development of next-generation medical devices for Industry 5.0. Biomed Eng Online. 2023;22(1):3.
50. Cheng J, Li W. Advancements in optical fiber communication technologies for Industry 5.0 applications. IEEE Commun Mag. 2020;58(5):48–54.
51. Bikowski J, Kowalski A. The impact of photonic technologies on the evolution of optical fiber communication systems in smart factories. J Lightwave Technol. 2021;39(24):7800–11.
52. Zhang Y, Wang X, Liu J. Optical fiber communication systems for data-intensive applications in Industry 5.0. Opt Express. 2022;30(14):24114–24.

# Chapter 6
# Real-Time Monitoring and Quality Control Using Photonic Technologies in Industrial Environments

**Satya Pratap Singh, Than Singh Saini, and Umesh Kumar Tiwari**

**Abstract** This chapter reviews the pivotal role of photonic technologies in real-time monitoring and quality control in industrial environments. It provides an overview of different types of photonic sensors, such as fiber optic temperature and strain sensors, photonic crystal fibers for gas sensing, fiber grating sensors and surface plasmon resonance biosensors. The chapter explores the fundamental principles of these sensors, their benefits and how they are used in different industrial applications. Moreover, the chapter covers various technologies based on laser and optical spectroscopy for the purposes of quality control in manufacturing. The chapter concludes with a discussion of existing challenges in implementing photonic technologies, such as interoperability, workforce development and high initial investment, and the promise of photonic technologies for greater sustainability, safety and efficiency in the next revolution of industrial automatization.

**Keywords** Photonics crystal fiber · Optical fiber · Sensors

## 6.1 Introduction to Real-Time Monitoring and Quality Control in Industry 5.0

The evolution of industrial automation has progressed through several stages, where each stage is characterized by their distinct technological advancements with the objective to enhance productivity and efficiency. Looking back, it is amazing how far industry has evolved, from the first Industrial Revolution (mechanization), through the Second (mass production), to the Third (automation) and Industry 4.0 had displayed a paradigm shift toward smart factories where physical processes are integrated with digital systems. However, Industry 5.0 shows an advanced stage

S. P. Singh · U. K. Tiwari (✉)
CSIR-CSIO, Sector 30, Chandigarh, India
e-mail: umeshtiwari@csio.res.in

T. S. Saini
National Institute of Technology Kurukshetra, Kurukshetra, Haryana, India



V. Bhardwaj et al. (eds.), *Photonics and Optoelectronics in Industry 5.0*, Advances in Optics and Optoelectronics, https://doi.org/10.1007/978-981-96-7594-4_6

where technological advancements fuse with social and ecological considerations, which ensures long-term development along with enhanced productivity. At its core, Industry 5.0 is about real-time monitoring and quality control to enable seamless operation, improved decision-making and proactive problem-solving within the manufacturing ecosystem.

As Industry 5.0 has real-time monitoring in which manufacturing operations can be continuously surveilled. It includes collecting data on every phase of the production process, from the raw material to the final product, using modern sensors and automated systems. With real-time data, manufacturers can identify inefficiencies, predict failures and optimize their operations. Such systems can be enhanced with additional sensors (temperature /humidity/fire detection) to monitor safety and compliance in an industrial setting [1]. Quality control is another critical component of Industry 5.0, focusing on precision and consistency across production workflows. Traditional quality control methods typically involve manual inspections or periodic sampling techniques, which can miss subtle defects. This is why new developments in modern real-time monitoring systems combine optical inspection techniques with machine learning algorithms to produce constant, high-quality output. Industry 5.0 brings forth better quality checks, one of them is automated optical inspection (AOI). AOI systems utilize sophisticated cameras and advanced imaging methods to detect potential defects that are missed by human. They can work around the clock without tiring or becoming distracted, promoting consistent, high-quality production. Moreover, the use of IoT (Internet of Things) technologies adds another layer to real-time monitoring in Industry 5.0. The interconnectivity of a number of previous individual technologies allows manufacturers to gather data from multiple sources at the same time, which is significant for deep analysis and data-driven decision-making [2]. This interconnectedness allows for predictive maintenance, where equipment failures are predicted proactively. One of the most notable features of Industry 5.0 is the introduction of smart robotics and AI (artificial intelligence) into quality control processes. Data coming from AOI systems can be utilized by machine learning algorithms, enabling these robots to adapt to new situations that could potentially increase their efficiency over time. But moving from classical manufacturing approaches to Industry 5.0 involves an enormous investment in client and server infrastructure—both hardware and software. This also means retrofitting existing facilities with modern sensors, IoT devices and AI-driven systems. Also, one has to train the workers to ensure that these new technologies are used properly.

The goal of this chapter is to present a holistic overview of impact of Industry 5.0 on the traditional manufacturing practices with the introduction of monitoring and quality control practices in real time, and the utilization of existing production lines by implementing AOI, machine learning algorithms and interconnected networks to reach unprecedented levels of efficiency, productivity and product quality. Real-time monitoring and management of production processes will be even more instrumental in staying competitive and reaching sustainable growth.

## 6.2 Photonic Sensors for Real-Time Monitoring

### *6.2.1 Fiber Optic Temperature Sensors*

Because of real-time monitoring applications, fiber optic temperature sensors have been established as an advanced media especially for the petrochemicals, aerospace and healthcare industries. These sensors use optical fibers to measure temperature with high accuracy over a long distance while being contactless and with minimal heat loss [3].

It is known that fiber optic temperature sensing utilizes light that interacts with the core of the optical fiber. Sending a laser beam through the fiber causes its intensity to drop due to thermal absorption and scattering when it crosses temperature gradients. By examining these fluctuations in signal intensity, one can measure temperatures precisely without any mechanical contact or interference. Fiber optic sensors are embedded with numerous benefits such as minimal loss (long-distance transmission), high sensitivity and fast response to the variation in parameters that make fiber optic sensor capable of real-time monitoring. Hence, they can be used for applications where a classical thermocouple can fail because of mechanical limitation [4]. Moreover, these sensors will not fall prey to electromagnetic interference and can survive in harsher conditions that would ruin traditional temperature sensing devices in no time. In addition, optical fiber technology can achieve multiple sensing points on one optical fiber, realizing distributed monitoring of its complex industrial systems. This multiplexed capability improves the overall efficiency of real-time sensing systems and increases the number of datasets by acquiring data from several locations. The need for continuous process control, such as in petrochemicals and refineries where safety, productivity and loss prevention are paramount, is ideally suited to the use of fiber optic sensors which can monitor temperatures in pipelines, and/or storage tanks based on a noncontact methodology. Fiber optics can provide accurate readouts for aerospace applications—where equipment must be insulated against shock waves; therefore, understanding the temperature at any point in time is critical, and fiber optics directly address this requirement while remaining noninvasive. In general, the use of fiber optic technology with real-time temperature monitoring provides substantial advantages in enhancing reliability, accuracy and operational efficiency across a multitude of industries. As shown in Fig. 6.1a, Favero et al. have experimentally demonstrated a PCF with germanium core in 2013. This PCF with a germanium core acts as an interferometer when sandwiched between two single-mode fibers for the temperature measurement. Figure 6.1b shows the experimental setup for the temperature sensing. In order to work in the reflection mode, the gold film was deposited to the end of the fiber. The proposed device exhibits an extremely high-temperature sensitivity of 78 pm/°C [3].

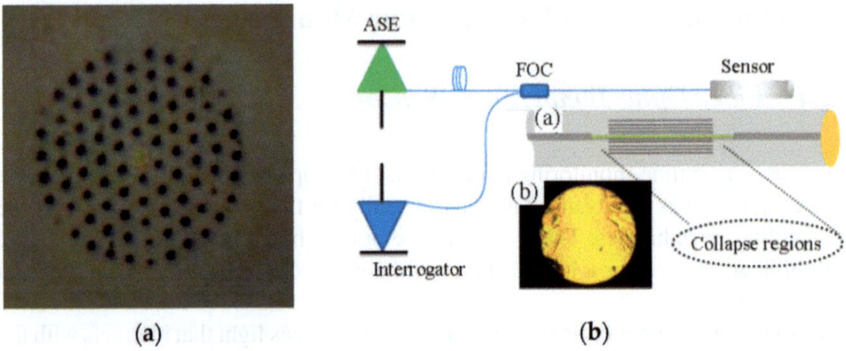


**Fig. 6.1** **a** Cross-section of the microstructured fiber and **b** experimental setup showing the collapsed region of the microstructured fiber with one end of the single-mode fiber gold-coated (adapted from Ref. [3] under the Creative Commons Attribution (CC-BY) license)

## *6.2.2 Fiber Optic Strain Sensing*

To address the need for accurate and real-time monitoring of strain in structures, fiber optics strain sensing has emerged as an important enabling technology with safety and operational improvements across a wide range of industrial applications. These advanced sensing systems use the special features of fiber optics, including immunity to electromagnetic interference and high spatial resolution, to yield real-time measurements of physical values like structural health and environmental conditions. Monitoring of cement sheath with distributed acoustic sensing (DAS) marks a breakthrough in well integrity management. Fiber optic sensors are integrated into the wellbore during the cement hydration process to consistently monitor acoustic and strain measurements along the cement column. These sensors allow for the identification of mechanical failures (which lies in cement sheath) on the real-time basis, as they are able to detect millimeter differences in the mechanical properties of cement sheath. This kind of DAS technology is valuable since it delivers a longitudinal image of the entire wellbore compared to traditional cement bond logs, which only provide discrete data points. When mechanical failures are identified, operators can proceed to take corrective actions to prevent those from turning into larger problems. This feature enables DAS systems to use deep well applications in high-temperature and high-pressure environments. Moreover, the specific localization of failure zones, made possible by the distributed nature of this system, helps in targeted remediation and optimization of resources [5, 6]. That being said, distributed fiber optic sensing can also be used to provide high-resolution strain monitoring for wind turbine gearboxes to detect failure on the fly. Researchers indicated that this type of system shows linear characteristics between load torque and strain measurements. Wind turbine blades with quasi-continuous fiber optic strain sensors allow for precise measurements of wind-induced strain and are critical for assessing structural integrity as well as optimizing their design. In these systems, optical frequency domain reflectometry (OFDR) helped maintain the accuracy of strain data, which simplified maintenance and safety assessments [7].

Another type of sensing technique is the use of single-mode–multimode–single-mode (SMS)-based fiber optic sensor for health monitoring of sandwich composites used in aerospace, automotive and construction sectors to detect delamination related failures. These sensors are capable of real-time detection of interfacial delamination. This allows for early detection and avoidance of failure in structural components. The potential application of this technology is to increase the safety, reliability, integrity and availability of vital infrastructure and equipment [8]. A novel label-free fiber optic biosensor based on a single-mode–tapered multimode–single-mode structure is presented for the detection of Salmonella Typhimurium. The sensor cladding was peeled off and further functionalized with anti-Salmonella Typhimurium antibodies, and subsequently, it was confirmed for detection in phosphate buffer saline and spiked milk samples. The sensor has a good sensitivity with the detection range (500–5000 CFU/mL) and limit of detection (LOD) (247 CFU/mL) with each analysis being completed in under 20 min. The sensor also showed superior selectivity toward *E. coli* and *S. aureus*. This biosensing platform proves to be rapid and specific in detection of pathogens due to its high sensitivity (275.86 nm/RIU [9].

A simple fiber Bragg grating (FBG) sensor could be designed for simultaneous measurement of strain and temperature since the FBG reflection wavelength has different sensitivities for strain and temperature. The sensor can distinguish accurately tensile transformations and temperature transformations by analyzing the shifts of the FBG reflection wavelength. This has huge applications in structural health monitoring and smart sensing systems [10].

### 6.2.3 Photonic Crystal Fibers for Gas Sensing

Photonic crystal fibers (PCFs) are a new generation of optical fibers that use the light around them to determine the surrounding environment. In contrast to conventional fibers that transmit light through a solid core, PCFs have a carefully designed array of air holes that run along the entire length of the fiber. These air holes can be arranged in a periodic or random pattern and create a unique photonic structure that controls the propagation of light through the PCF. The resultant light confinement within the PCF core and the generation of photonic bandgaps due to its specific structure makes them good candidate for sensing applications. As shown in Fig. 6.2, it can be seen that in the low confinement loss bandwidth, the relative sensitivity of the optimized PCF is greater than 95.70% [11]. The core principle behind PCF gas sensing lies in the ability of fiber to detect changes in the refractive index of the surrounding environment. When a gas molecule enters the PCF, it interacts with the light propagating through the core of the PCF. This interaction alters the effective refractive index of the PCF that results in shift in the wavelength of the transmitted light. By carefully analyzing this wavelength shift, one can identify the specific gas and its concentration. PCFs offer several advantages over traditional gas sensors that

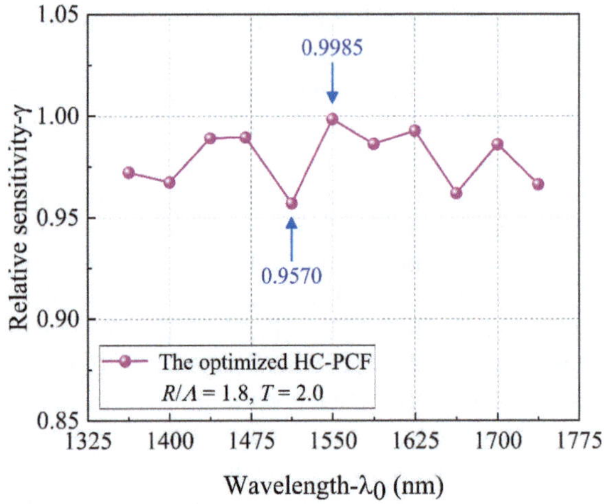


**Fig. 6.2** The relative sensitivity of the optimized PCF in the range of the low confinement loss bandwidth (adapted from Ref. [11] under the Creative Commons Attribution (CC-BY) license)

make them a compelling choice for real-time monitoring in industrial environment. PCFs are flexible and robust, and the unique structure of PCFs allows for a significant interaction between light and the surrounding gas. This leads to high sensitivity and accurate detection of even low gas concentrations [12]. By carefully designing the PCF structure, one can achieve simultaneous detection of several gases. PCFs are effective for detecting low concentrations of dangerous gases, which is crucial for safety inspections in industrial settings [13].

### 6.2.4 *Fiber Grating Sensors*

Fiber grating sensors are being applied in medicine to monitor vital signs such as blood pressure, temperature and oxygen saturation in real time, enabling early detection and treatment of various medical conditions. Additionally, fiber grating sensors are being used to track the movement and position of surgical instruments and implants. This will helpful in improving surgical precision and reducing complications [14]. Tiwari et al. have experimentally demonstrated a fiber Bragg grating-based compact and high-sensitive single-axis accelerometer. With the help of simulations, its performance with cantilevers of various materials is compared. The accelerometer was tested with random and known seismic vibrations. This sensor shows a good response to signals like footsteps and hammering. It achieved a sensitivity of 88 pm/g over a broad bandwidth of 5 to 100 Hz. This fully packaged device is suitable for monitoring vibrations in civil structures like highways, bridges, dams and aeronautical platforms [15].

### 6.2.5 Surface Plasmon Resonance Biosensors

Surface Plasmon Resonance (SPR) sensors are increasingly recognized for their potential in real-time monitoring within the medical industry. These sensors exploit the interaction of light and surface plasmons, providing a label-free approach to detect biomolecular interactions, suitable for point-of-care diagnostics. SPR sensors demonstrate real-time detection of biomolecules interaction, which is an essential prerequisite for timely medical diagnosis [16]. By allowing observation of interactions without labels, it minimizes complexity and possible interference, increasing sensitivity [17]. Greater than 20,000 nm/RIU sensitivity has been established with some SPR configurations, facilitating detection for low-concentration analytes. For example, SPR sensors detect specific biomolecules, such as antibodies and proteins, and can be used for disease diagnosis and monitoring [18, 19]. Researchers have developed a new plasmon-based platform for real-time monitoring of proteolytic activity that illustrates potential applications in bleeding disorder diagnostics [20]. Similarly, a tilted fiber Bragg grating (TFBG) SPR sensor-based system that measures pH in the range of 2–12 using a thin layer of polyaniline was introduced in [21] and has a relatively high sensitivity of 0.50335 nm/pH. Although sensing experiments using SPR are very promising, the following challenges are in the SPR sensors, such as robust calibration, interference from complex biological samples, sensor stability and complex fabrication processes, which hinders its widespread application in medical industry. Nonetheless, ongoing advancements in sensor technology continue to enhance their applicability in real-time medical monitoring.

Kaushik et al. have demonstrated a fiber optic SPR immunosensor with $MoS_2$ nanosheets for sensitive detection of *E. coli*. The nanosheets, prepared by chemical exfoliation, were used to immobilize *E. coli* antibodies, enabling rapid (~15 min) bacterial detection with high specificity. The sensor showed a strong linear response ($R^2 = 0.994$) and achieved a low detection limit of 94 CFU/mL with high sensitivity (3135 nm/RIU). It performed well in spiked water and juice samples, making it ideal for monitoring water and food quality for pathogens [22].

Because of their high accuracy for detecting biomolecules, surface plasmon resonance techniques are useful in SARS-CoV-2 detection. Traditional RT-PCR tests are reliable but time-consuming and costly. SPR biosensing chips offer fast, accurate and portable detection; these characteristics make them ideal for COVID-19 diagnosis. These innovative biosensors are promising tools for combating the pandemic with quick and reliable results. Figure 6.3 illustrates the proposed SPR sensor designed for the rapid detection of the SARS-CoV-2 virus. The sensor has a five-layer sensing region structure based on the Kretschmann–Raether configuration [23, 24].

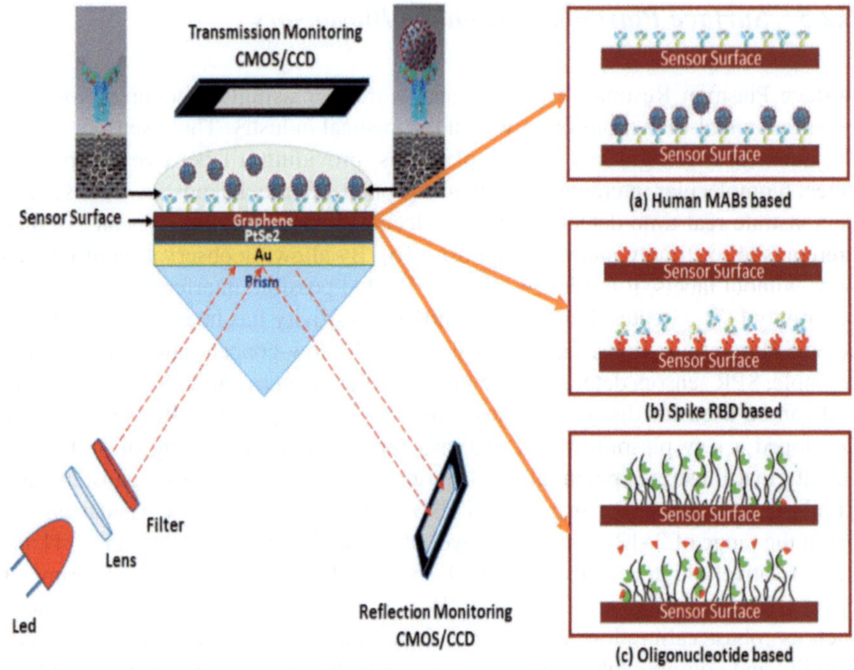


**Fig. 6.3** A schematic representation of the five-layered (Bk7/Au/PtSe2/Graphene/PBS) SPR biosensor designed for detecting SARS-CoV-2 viruses is presented. The sensor operates in three detection modes: **a** rapid detection of the entire viral spike using human mAbs immobilized on the viral spike RBD (spike RBD as the analyte and mAbs as the ligand), **b** swift mAbs recognition facilitated by immobilized viral spike RBD (mAbs as the analyte and spike RBD as the ligand) and **c** detection of viral RNA sequences via the graphene-embedded sensor surface with immobilized probe sequences (adapted from Ref. [24] under the Creative Commons Attribution (CC-BY) license)

## 6.2.6 Photonic Integrated Circuit-Based Sensors

PICs have been proposed as key devices for the continuous and real-time monitoring in industrial context, such as for gas sensing and public safety monitoring in environmental control. PICs-based sensors allow for high sensitivity and fast response times thanks to advanced photonic technologies—for example, waveguides and resonators. This characteristic of PICs-based sensors makes them better compared to the traditional sensors [25, 26]. These sensors can be linked to various monitoring systems in industries, improving safety standards and increasing performance [1]. In PICs, microring resonator (MRR)-based gas sensors are emerging as highly sensitive and compact devices for detecting various gases. These sensors rely on the changes in refractive index due to interactions of gases, utilizing the concept of resonant wavelength change. Igarashi et al. showed a silicon MRR sensor based on a Pt–$SiO_2$ thin

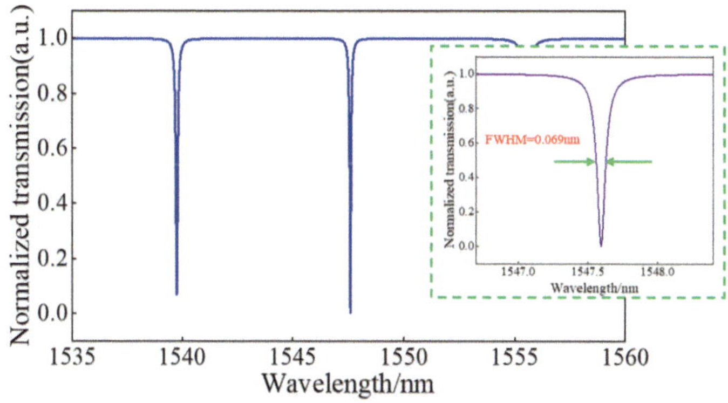


**Fig. 6.4** Normalized transmission spectrum of the MRR with the deionized water cladding. The illustration in the green dashed box shows the full width at half height of the main resonant peak more clearly (adapted from Ref. [29] under the Creative Commons Attribution (CC-BY) license)

film for hydrogen detection [27]. Hydrogen levels are measured based on the resonant wavelength shift attributed to heat in the Pt–$SiO_2$ film. Shifts of 5.0 nm and 2.4 nm were detected at 4.0% and 0.4% hydrogen concentrations, respectively. Its sensitivity is further improved by the inclusion of an $Al_2O_3$ cladding layer.

MRR sensors are suitable for lab-on-chip applications, and with the help of machine learning techniques, they become capable of classifying and detecting multiple gases simultaneously. The integration of advanced materials could further improve the performance of MRR-based sensors in environmental monitoring [28]. A novel hybrid waveguide double-slot subwavelength grating MRR sensor is proposed for water refractive index environment. Figure 6.4 shows the initial normalized transmission spectrum of the sensing device exposed under the deionized water layer, with three resonant peaks located at 1539.74 nm, 1547.59 nm and 1555.52 nm, respectively. This sensor achieves a sensitivity of 1005 nm/RIU, 182 nm/RIU higher than current designs, along with a high Q factor of 22,429. This sensor has a low detection limit of $6.86 \times 10^{-5}$ RIU. Its improved design enhances analyte–photon interactions that makes it highly effective for sensing applications [29] (Table 6.1).

## 6.3 Optical Techniques for Quality Control

### *6.3.1 Optical Methods for Quality Control Monitoring in Manufacturing Processes*

Quality control is essential for maintaining high standards in many industries. It ensures products are reliable, safe and efficient. In today's competitive market, strong quality control practices are more important than ever. In this part of the chapter, we

**Table 6.1** Various photonic sensors for real-time monitoring

| Sensor type | Features | Applications |
|---|---|---|
| Fiber optic temperature sensors [3, 4] | – High precision over long distances<br>– Resistant to electromagnetic interference<br>– Operates in harsh environments<br>– Multiplexing capability for distributed monitoring<br>– Noncontact temperature measurement | Petrochemicals (pipeline and storage tank monitoring)<br>Aerospace (robust temperature monitoring)<br>Health care (real-time temperature monitoring) |
| Fiber optic strain sensing [5–10] | – High spatial resolution<br>– Resistance to electromagnetic interference<br>– Distributed acoustic sensing (DAS) for real-time monitoring<br>– Operates in high-temperature and high-pressure environments | Cement sheath monitoring in wellbores<br>Wind turbine gearbox monitoring<br>Structural integrity assessment in aerospace, automotive and construction |
| Photonic crystal fibers (PCFs) [11–13] | – High sensitivity to refractive index changes<br>– Flexible and robust<br>– Simultaneous detection of multiple gases<br>– High sensitivity to low gas concentrations | Industrial gas sensing (safety inspections)<br>Environmental monitoring (low-concentration gas detection) |
| Fiber grating sensors [14, 15] | – Compact and high-sensitive<br>– Measures strain and temperature simultaneously<br>– High sensitivity to vibrations | Civil structure monitoring (bridges, dams, highways)<br>Aeronautical platforms<br>Medical applications (vital sign monitoring, surgical instrument tracking) |
| Surface plasmon resonance (SPR) [16–24] | – Label-free biomolecular detection<br>– High sensitivity (> 20,000 nm/RIU)<br>– Real-time feedback<br>– Rapid detection (~ 15 min) | Medical diagnostics (disease detection, proteolytic activity monitoring)<br>Pathogen detection (*E. coli*, SARS-CoV-2)<br>Water and food quality monitoring |
| Photonic integrated circuits (PICs) [1, 25–29] | – High sensitivity and rapid response<br>– Compact design<br>– Integration with machine learning for gas classification<br>– Enhanced analyte–photon interactions | Industrial gas detection (hydrogen sensing)<br>Environmental monitoring (water refractive index sensing)<br>Lab-on-chip applications |

explore how photonic technologies can transform quality control that have potential to offer precise, real-time insights to improve processes and ensure product quality. In manufacturing, it is important to ensure products meet set standards. This keeps customers happy, follows regulations and supports business success. Poor quality control can lead to defects, higher costs from rework or recalls, and harm to the brand. Industries like pharmaceuticals, electronics, automotive, food processing and environmental monitoring face tough challenges in maintaining quality while meeting strict rules. Photonic technologies involve the use of light (photons) for various applications. These include optical sensors [30–32], spectroscopy [33], imaging systems [34, 35] and communication devices. In industrial settings, photonic tools can provide real-time data that enable continuous monitoring and rapid intervention when deviations from standards are detected [36, 37]. The use of photonic technologies in quality control offers several advantages. Firstly, these tools are noninvasive and can operate continuously without disrupting the production line. Secondly, they provide real-time data that allow for immediate corrective actions, reducing downtime and increasing productivity. Additionally, photonic systems can be integrated with advanced analytics and machine learning algorithms to predict potential issues before they occur [38, 39].

Optical techniques offer enriched technology for quality control monitoring in different areas including pharmaceutical, chemical and materials [40]. The laser technologies afford a fundamental tool in the optical metrology, and fiber optics deals resourcefulness for multiparameter measurements.

The optical industries and the various method that are being employed in optical techniques for quality control are given in Table 6.2.

This section aims to introduce high-resolution optical techniques developed for the quality control in manufacturing. Most of the such techniques are based on the fiber optic sensors that have been employed in the previous several years as a feasible alternative to conventional methods for quality monitoring in the industrial processes.

As we know, the energy of photon is given by the relation: $E = h\nu$; here, $h$ is the plank's constant, and $\nu$ is the frequency. This formula indicates that the shorter wavelengths have more energy than the electromagnetic spectrum with larger wavelengths. Therefore, the interaction of ultraviolet (UV) light with the matters can produce some optical phenomena including fluorescence which can deliver applicable information about the sample.

**Table 6.2** Various optical industries and methods

| Industry | Methods |
|---|---|
| Industry 4.0 | – UV–VIS Spectroscopy<br>– Raman<br>– Photoluminescence<br>– FTIR |
| Materials | Diffuse reflectance |
| Biopharmaceutical | Colorimetry |
| Food products | Absorbance |

In the optical metrology, visible light is extensively used since its versatility. Visible light can be used in the economical UV–Vis spectrometers to analyze transmittance, reflectance and absorbance. Further, it can be useful in the quality inspection in the diverse applications including manufacturing process of the solar cells and elaboration of food products or the manufacturing process of solar cells. If we talk about the lasers in the visible and near-IR regions, they are being used in the chemical identification based on the Raman scattering technique. Currently, the pharmaceutical industry performs quality control measures based on the Raman scattering. Furthermore, infrared light is being used in the optoelectronic devices as a transmitter and receiver end. Fourier transform infrared (FTIR) spectroscopy uses broadband infrared light extending from 2 to 20 $\mu$m (wavenumbers: 500–5000 $cm^{-1}$). FTIR technique is used particularly for the identification of the functional groups in the organic and inorganic substances.

## 6.4 Noncontact Optical Methods for Quality Monitoring

There is always better to have interaction with material without physical contact. The optical techniques for monitoring manufacturing processes provide the ability to take measurements without physical contact with the sample.

### *6.4.1 Automobile Optical Inspection*

Trained manpower is frequently employed to carry out quality inspections in industrial processes. However, human examination has limitations, for example, low operational speeds and the need for regular rotation of personnel. To address these challenges, optical techniques have emerged as a viable alternative for automating the inspection of automotive components. In the automotive industry, automated visual inspection methods have recently been proposed for the assessment of head-up displays (HUDs) [41]. These devices do not have any moving parts and are made especially for checking regular patterns. The inspection procedure assures correct signal representation without distortion by using Fourier filtering and operating below the Shannon–Nyquist criterion. Notably, this approach does not require high-resolution cameras or scanning equipment, which greatly reduces the cost and improves the efficiency of the inspection process.

Vibration measurements for quality control in automobile industry often utilize acoustic waves produced by vibrations to determine the working conditions in internal combustion engines. However, systems based on acoustic techniques have limitations—an important drawback being the need for a sound-isolated environment to eliminate interference. This limitation is based on urgency which can reduce productivity during inspections as well.

Optical vibration measurement is reliable as it is not affected by other acoustic sources, and it allows for in situ measurements. This technique takes advantage of the laser Doppler effect, which detects frequency shifts of light waves relative to the same shifts of vibration. The use of this principle enables more accurate and noninvasively optical methods to monitor vibrations. By using this optical method, the efficiency and accuracy of quality inspections in automotive manufacturing is enhanced.

### 6.4.2 Materials

Materials play a pivotal role in manufacturing and engineering due to their electrical, thermal and mechanical properties. These features of materials directly influence the product quality. Cement, being one of the most extensively used materials in civil engineering, exemplifies this significance. Its physical attributes, e.g., density, strength and consistency, are greatly influenced by the method of fabrication. However, many fabrication techniques remain empirical. Therefore, there is limited information about the fabrication technique and that results in uncertain outcomes.

To address these challenges, optical techniques have been explored to study and characterize the physical properties of cement. One notable optical method is diffuse reflection, which can be used to study the hydration process in cement paste [42]. When water is added to cement during the setting process, then heat is generated due to hydration reactions that lead to a change in temperature. According to the Powers–Brunauer model, these temperature changes are closely linked to the chemical processes occurring at different stages of the hydration process in cement paste. Researchers have observed that the diffuse reflectance pattern closely mirrors the temperature changes during hydration when analyzed using a He–Ne laser with an emission wavelength of 633 nm. This correlation provides valuable insights into the hydration behavior of cement. As illustrated in Fig. 6.5, this optical technique offers a reliable method for analyzing the setting process that enables better understanding and control of the material properties.

Complex elasticity module has a fundamental importance in asphalt durability. Currently, there are standard methods to determine complex elasticity, and however, conventional approaches require contact devices, e.g., accelerometers or piezoelectric devices, and are limited to low frequencies (30 Hz). To overcome these limitations, an optical method, Timoshenko's beam theory, which is an advanced theory used in structural engineering and mechanics to describe the mechanical properties is used [43]. This method utilizes a scanning laser Doppler vibrometer combined with an identification framework. The technique eliminates the need for direct contact with the material that makes it a noninvasive and efficient alternative for evaluating complex elasticity module. By integrating optical measurement systems, it becomes possible to achieve accurate and high-frequency assessments of complex elasticity that is helpful in the study of asphalt durability.

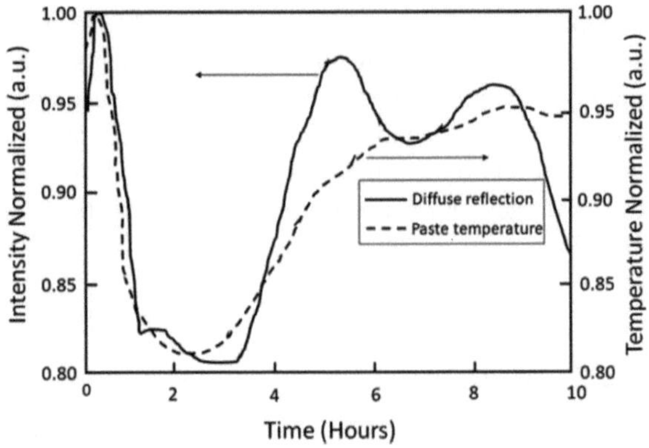


**Fig. 6.5** Temperature versus reflectance on the setting of cement paste (adapted from Ref. [42] under the Creative Commons Attribution (CC-BY) license)

(a) **Laser in Manufacturing**

Lasers possess unique features, such as monochromaticity, high directionality and coherence. These characteristics make them incredibly versatile in manufacturing industry. Their applications in manufacturing can be categorized into two main areas, e.g., material processing and precision measurement. Material processing includes tasks like drilling, cutting, welding and marking, but lasers can also play a critical role in specialized applications such as surface treatments.

Optical metrology involves technologies like 3D laser scanners, free-space optical detectors and fiber optics sensors, etc. These technologies are particularly valuable for industries, such as aerospace and automotive manufacturing, that require precise measurements. The integration of automated measurements with manufacturing processes ensures high-quality production standards. Among the above various optical measurement tools, the 3D laser scanner stands out for its efficiency and accuracy. This technology offers numerous advantages, such as rapid operation, superior resolution on the scale of microns and user-friendly handling. Its ability to perform noncontact dimensional measurements makes it an essential tool for industries that demand precise and reliable inspections. By leveraging laser-based technologies, manufacturers can achieve higher efficiency, improved quality control and enhanced product performance.

## 6.5 Challenges and Future Directions

While the amalgamation of photonic technologies with Industry 5.0 offers numerous benefits, there are many challenges that must be overcome for the integration of photonic technologies with Industry 5.0 to be truly effective.

(b) **High Initial Investment**

The implementation of photonic technologies requires a large amount of CAPEX in hardware, software and infrastructure. The use of modern sensors and devices for

IoT and AI-based systems can be costly for any small/medium enterprise (SMEs) to upgrade existing facilities or new setup. In Industry 5.0, the government and the photonic industry stakeholders should make sure the financial and supportive help in providing these photonic technologies.

(c) **Workforce Training**

A workforce that can operate and maintain these systems will be essential to the successful integration of photonic technologies into Industry 5.0. However, training programs and educational initiatives are required for the workers to attain the skills and knowledge to familiarize themselves with photonic technologies. The development of a photonics-proficient workforce requires collaboration among academia, industry and government.

(d) **Interoperability and Standardization**

Existing industrial systems have dependencies on data and communications interfaces. So, the integration of photonic technologies with existing systems demands interoperability and standardization. Interoperability between the various sensors, devices and platforms is essential for the success of Industry 5.0. Standards and protocols for photonic technologies development will enable straightforward integration and adoption.

(e) **Data Security and Privacy**

If photonic technologies are extensively utilized in Industry 5.0, it can result in a huge amount of data; however, this raises potential issues regarding data security and privacy. It necessitates high-level cybersecurity to protect sensitive data and data protection regulations to develop trust in the photonic technologies. Such challenges call for best cybersecurity practices and data governance frameworks.

The future of photonic technologies in Industry 5.0 depends on the development of more intricate and interconnected systems. The next generation of photonic sensors and devices will be powered by advancements in artificial intelligence, nanotechnology, quantum technology and other fields. For instance, ultra-sensitive sensors and secure communication systems may be made possible by the combination of Industry 5.0 with quantum photonics. Additionally, the creation of flexible and wearable photonic devices might result in novel uses for healthcare and human–machine interaction.

## References

1. Anal S, Kaur S. Live data monitoring in industry. Int J Res Appl Sci Eng Technol. 2024;12(6):304–13.
2. Industrial AT, System M. Int J Adv Res Sci Commun Technol. 2024;4(1):372–6.
3. Favero FC, Becker M, Spittel R, Rothhardt M, Kobelke J, Bartelt H. Micro-structured fiber interferometer as sensitive temperature sensor. Photonic Sensors. 2013;3(3):208–13.

4. Yang T, Shen T, Feng Y, Liu X. Application of microstructured fiber sensor in the field of temperature detection. Embed Selforganising Syst. 2022;9(3):8–11.
5. Amer R, Xue Z, Hashimoto T, Nagata T. Distributed fiber optic temperature and strain sensing in cementing and water injection: insights to well integrity monitoring and multisensing optical fiber cable design. Gas Sci Eng. 2024;130: 205430.
6. Wang S, Yang Z, Mohanty L, Zhao C, Han C, Li B, et al. Distributed fiber optic sensing for internal strain monitoring in full life cycle of concrete slabs with BOFDA technology. Eng Struct. 2024;305: 117798.
7. Knebusch J, Gundlach J, Meier D, Meier T, Govers Y. Static testing using OFDR-based quasi continuous fiber optic strain sensing on the WiValdi wind turbine rotor blade. J Phys Conf Ser. 2024;2767(4).
8. Mitra N, Patra AK, Singh SP, Mondal S, Datta PK, Varshney SK. Interfacial delamination in glass-fiber/polymer-foam-core sandwich composites using singlemode–multimode–single-mode optical fiber sensors: identification based on experimental investigation. J Sandw Struct Mater. 2020;22(1):40–54.
9. Kaushik S, Pandey A, Tiwari UK, Sinha RK. A label-free fiber optic biosensor for Salmonella Typhimurium detection. Opt Fiber Technol. 2018;46:95–103.
10. Reddy PS, Prasad RS, Gupta D Sen, Shankar MS, Srimannarayana K, Tiwari U, et al. A simple FBG sensor for strain–temperature discrimination. Microw Opt Technol Lett. 2011;53(5):1021–4.
11. Wan B, Zhu L, Ma X, Li T, Zhang J. Characteristic analysis and structural design of hollow-core photonic crystal fibers with band gap cladding structures. Sensors. 2021;21(1):284.
12. Zheng L, Liu Y. Characteristic analysis and structural design of photonic crystal fibers for gas sensing applications. Preprints. 2023.
13. Wang G, Sun J, Li T, Wang H, Li J. Multiplexed photonic crystal fiber gas-sensing network based on intracavity absorption. Sensors. 2022;22(23):9237.
14. Mishra V, Singh N, Tiwari U, Kapur P. Fiber grating sensors in medicine: current and emerging applications. Sensors Actuators A Phys. 2011;167(2):279–90.
15. Tiwari UK, Saini TS, Negi V, Pandey A, Garg H, Sinha RK. Design and development of a field deployable packaged fiber Bragg grating-based accelerometer. Opt Eng. 2019;58(1):014104–014104.
16. Das CM, Yang F, Yang Z, Liu X, Hoang QT, Xu Z, et al. Computational modeling for intelligent surface plasmon resonance sensor design and experimental schemes for real-time plasmonic biosensing: a review. Adv Theory Simulations. 2023;6(9):1–28.
17. Sayin S, Zhou Y, Wang S, Acosta Rodriguez A, Zaghloul M. Development of liquid-phase plasmonic sensor platforms for prospective biomedical applications. Sensors. 2024;24(1):186.
18. Vashistha N, Abuleil MJ, Shrivastav AM, Bajaj A, Abdulhalim I. Real-time ellipsometric surface plasmon resonance sensor using polarization camera may provide the ultimate detection limit. Biosensors. 2023;13(2):173.
19. Liu W, Liu C, Wang J, Lv J, Lv Y, Yang L, et al. Surface plasmon resonance sensor composed of microstructured optical fibers for monitoring of external and internal environments in biological and environmental sensing. Results Phys. 2023;47(106365):1–11.
20. Oliveira-Silva R, Wang Y, Nooteboom SW, Prazeres DMF, Paulo PMR, Zijlstra P. Single-particle plasmon sensor to monitor proteolytic activity in real time. ACS Appl Opt Mater. 2023;1(10):1661–9.
21. Huang Y, Wang E, Li P, Cao J, Lyu G. PH sensor based on tilted fiber Bragg grating surface plasmon resonance with a polyaniline reaction deposition film layer. Opt Quantum Electron. 2024;56(8):10887.
22. Kaushik S, Tiwari UK, Pal SS, Sinha RK. Rapid detection of Escherichia coli using fiber optic surface plasmon resonance immunosensor based on biofunctionalized Molybdenum disulfide ($MoS_2$) nanosheets. Biosens Bioelectron. 2019;126:501–9.
23. Pandey PS, Raghuwanshi SK, Shadab A, Ansari MTI, Tiwari UK, Kumar S. SPR based biosensing chip for COVID-19 diagnosis—a review. IEEE Sens J. 2022;22(14):13800–10.

24. Akib TBA, Mou SF, Rahman MM, Rana MM, Islam MR, Mehedi IM, et al. Design and numerical analysis of a graphene-coated SPR biosensor for rapid detection of the novel coronavirus. Sensors. 2021;21(10):3491.
25. Butt MA, Piramidowicz R. Integrated photonic sensors for the detection of toxic gasses—a review. Chemosensors. 2024;12(7):143.
26. Williams K, Latkowski S, Moskalenko V, Llorens-Revull M, Bente E. Optical sensing using InP integrated photonics. In: Progress in electromagnetic research symposium (PIERS). 2016. p. 2681.
27. Igarashi J, Okazaki S, Nishijima Y, Higo A, Arakawa T. Characteristics of highly sensitive hydrogen gas sensor based on Pt-$SiO_2$/Si microring resonator. Jpn J Appl Phys. 2024;63(5):05SP17.
28. Malathi S, Roy UM, S SK, Tiwari U, Mishra S, Gupta S, et al. Optimization of integrated optic micro ring resonator gas sensor using machine learning technique. In: Proceedings SPIE 12638, women in optics and photonics in India 2022. 2023. p. 126380I.
29. Lu K, Huang B, Lv X, Zhang Z, Ma Z. Ultrasensitive silicon photonic refractive index sensor based on hybrid double slot subwavelength grating microring resonator. Sensors. 2024;24(6):1929.
30. Gritsenko T V., Chesnokov GY, Koshelev KI, Khan RI, Stepanov K V., Valba O V., et al. Optical fiber sensor for real-time monitoring of industrial structures and application to urban telecommunication networks. In: 2023 IEEE international conference on metrology for extended reality, artificial intelligence and neural engineering, MetroXRAINE 2023—Proceedings. IEEE; 2023. p. 383–8.
31. Faraco RL, Barino F, Campos D, Sampaio G, Honório L, Marcato A, Bessa dos Santos A, Dos Santos CC, Hamaji F. Hydroelectric plant safety: real-time monitoring utilizing fiber-optic sensors. Sensors. 2024;24(14):4601.
32. Mendez A. Photonic sensing technology in the energy sector. In: Optics infobase conference papers. 2012. p. 3–5.
33. Xu F, Zhang Y, Somesfalean G, Wang H, Wu S, Zhang Z. Broadband spectroscopic sensor for real-time monitoring of industrial $SO_2$ emissions. Appl Opt. 2007;46(13):2503–6.
34. Sioma A. Vision system in product quality control systems. Appl Sci. 2023;13(2):751.
35. He H, Fischer C, Darsow U, Aguirre J, Ntziachristos V. Quality control in clinical raster-scan optoacoustic mesoscopy. Photoacoustics. 2024;35: 100582.
36. Ebayyeh AARMA, Mousavi A. A review and analysis of automatic optical inspection and quality monitoring methods in electronics industry. IEEE Access. 2020;8:183192–271.
37. Burmistrov DE, Pavkin DY, Khakimov AR, Ignatenko DN, Nikitin EA, Lednev VN, et al. Application of optical quality control technologies in the dairy industry: an overview. Photonics. 2021;8(12):551.
38. Karathanassi V, Iossifidis C, Rokos D. Application of machine vision techniques in the quality control of pharmaceutical solutions. Comput Ind. 1996;32(2):169–79.
39. Wiesinger F, Klepatsch D, Bogner M. Optical quality control using deep learning. In: Proceedings of the 31st European modeling & simulation symposium (EMSS 2019). 2019. p. 119–27.
40. Alcaraz JLG, Sánchez-Ramírez C, AJGL. Techniques, tools and methodologies applied to quality assurance in manufacturing. Springer International Publishing; 2021. 23 p.
41. Ferreira FP, Forte PM, Felgueiras PE, Bret BP, Belsley MS, Nunes-Pereira EJ. Automatic optical inspection of regular grid patterns with an inspection camera used below the Shannon-Nyquist criterion for optical resolution. In: Photonic instrumentation engineering IV. 2017. p. 419–26.
42. Martínez HD, Bernal JS, Mota RG R-CI. Optical evaluation on the setting of cement paste. J Phys Conf Ser. 2015;582(1):012021.
43. Navid Hasheminejad, Vuye C, Margaritis A, Van, Joris Dirckx, Vanlanduit S. Characterizing the complex modulus of asphalt concrete using a scanning laser doppler vibrometer. Materials (Basel). 2019;12(21):3542.

# Chapter 7
# Role of Photonics in Enhancing Additive Manufacturing Techniques and Process

**Ritu Goel and Ranjana Jha**

**Abstract** In the material science development sector, twentieth century seems to be a revolutionary era. With the innovative materials like semiconductors, biomaterials, polymers, etc. have been upgraded for the latest technologies. These achievements and new materials have diversified the applications at industrial levels such as in transportation, health care, electronics, and energy devices. Photonics-based additive manufacturing technology has ushered the process of 3D printing at a low-cost with lesser fabrication steps. It has also reduced the machinery size, material wastage, and is able to create more complex 3D patterns. In this chapter, we will discuss the types of photonics-based additive manufacturing techniques along with their benefits and the limitations. Also, we will relate a few parameters of this latest technology with the traditional techniques to manufacture 3D objects in an effective manner.

**Keywords** Additive manufacturing techniques · Photonics · Lasers · 3D printing · Stereolithography

## 7.1 Introduction to Additive Manufacturing and Photonics

### *7.1.1 Overview of Additive Manufacturing (AM) Techniques*

In 1983, an American engineer Chuck Hull at 3D systems corporation has designed a solid imaging process called stereolithography or the 3D printer [1]. He has also developed the first commercial rapid prototype technology with the STL file format. Additive manufacturing (AM), also known as 3D printing, causes a transition in the manufacturing sectors to become large-scale load bearing structures from micro-precision medical components [2].

AM techniques allow to have intricate designs to transforms and reshape the build products via 3D printing and results in a sustainable manufacturing environment.

R. Goel (✉) · R. Jha
Indira Gandhi Delhi Technical University for Women, New Delhi, India
e-mail: goelritu1991@gmail.com

 
V. Bhardwaj et al. (eds.), *Photonics and Optoelectronics in Industry 5.0*, Advances in Optics and Optoelectronics, https://doi.org/10.1007/978-981-96-7594-4_7

Thus, it significantly reduces the wastes and by-product generated by the traditional manufacturing approaches. AM technology works on the phenomenon of creating a 3D object by adding layer by layer of the material and fusing them through a CAD-generated digital model [3]. Thus, AM techniques provide flexibility in choosing a variety of materials with complex geometries product and results in lesser generation of waste.

As per the ISO/ASTM standards, shown in Fig. 7.1, AM techniques have been divided into seven categories depending upon the energy source used, type of layering of the material, and the fusing material [4].

1. **Vat Photo-Polymerization**

"Photo" means light and "polymerization" means the process of combining small molecules or monomers into a chainlike network structure called polymers. Thus, it is a technique to design a 3D structured solid object by selectively exposing the material under ultraviolet light. This technique is used only for printing plastics and producing high-resolution parts in industries like jewelry and dentist. Despite having significant benefits, Vat polymerization can only be done using radiation-curable raisins or photo-resin materials which requires substantial level of post-processing.

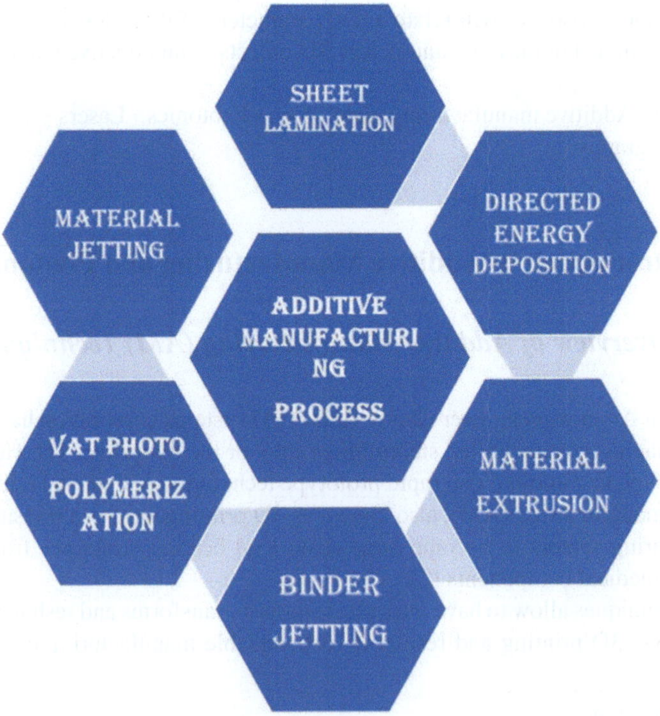


**Fig. 7.1** Types of additive manufacturing processes

2. **Material Extrusion**

Extrusion is a process of creating an object via pushing the required material through a desired cross-sectional nozzle which results in a structured shaped manner. Thus, material extrusion is a technique of printing a desired cross-sectional object by feeding the required material into the printer via coil. The material then gets heated at the nozzle tip and the melted material gets deposited layer by layer which further cools and solidify to form an object. Because of its low-cost equipment and easy handling, this technique has been used by many recreational users. But this technique is only limited with using plastic polymers-based materials as the heating elements cannot work with high-density metals used in fixturing and tooling.

3. **Material Jetting**

It is a technique of dropping or jetting liquid waxes or photopolymers selectively on to the target and curing it by heat or ultraviolet light which results in a 3D structured objects. In this technique, different types of materials can be used to create a fully colored, high-resolution 3D objects. Although, it is a time-consuming process as the target needs to build one droplet at a time and also it works only with wax-based materials which are fragile sometime.

4. **Binder Jetting**

Inkjet powder printing or Binder Jetting is the commonly used printing technique for 3D objects. In this technique, the printer head moves vertically and horizontally over the material to build a new layer of powder with every pass. A variety of material like glass, plastics, ceramics, gypsum powder, sand, etc., can be used to create multiple layers of a 3D object. Because of these types of material, it is a very economical and faster AM technique. But the only limitation is that the created 3D objects require cleaning and hardening as a post-processing to avoid being fragile.

5. **Powder Bed Fusion**

Powder bed fusion is the technique of using a powdered material spread in a large cross-sectional area or bed which undergoes selective fusion with the help of an electron or a laser beam. Thus, it is also called as an electron beam melting (EBM) technique. Powdered materials like ceramic powder mixed with sand, metal, plastic, etc., get selectively fused, the targeted area moves down to create a new layer upon it via the same process. Because of the high complexity level in printing the object, it results in a stable and a stronger 3D object [5]. However, it requires a large space to make a bed a powder of the material; it seems to be difficult to maintain a clean working environment for the process.

6. **Directed Energy Deposition (DED)**

DED is the frequently used technique for metal printing and to improve the production of individual parts. In this technique, multi-axis arm having a nozzle mounted upon it helps in depositing the melted metal wire or powder onto the target. The material has been melted using a focused laser or an electron beam as an energy source. So, when the liquid material precisely dropped on the target material, it quickly hardens to form a layer. This process continues repeating until the required printing has been finished. Also, the significant benefit of this technique is that it can help in depositing multiple materials within a single printing process.

7.**Sheet Lamination**

It is a technique of stacking thin sheets of the required material and bonding them together via ultrasonic bonding, welding or brazing. Due to this process, sheet lamination technique can also be called as ultrasonic AM or lamination object manufacturing. Thus, the object will result in a desired structure once all the layers are stacked and laminated properly and is further refined by a CNC machine which removes excess material. Sheet lamination is an economical process in which one can deposit layers of different material. But the technique is limited with the use of sheet structured materials like metal, plastic, paper, etc., and results in the wastage of the material while reefing the object at the end of the process.

### *7.1.2 Basics of Photonics and Its Key Components*

Photonics plays a vital role in the AM techniques. As discussed in the above techniques, LED, lasers, and other light sources help in curing, melting, and depositing every single layer in printing to form a 3D objects [6]. Photonics technology can also work as a spectral sensor. These types of sensors help in monitoring the process or it can also check the fault spots before and after the production.

## 7.2 Photonics-Based Additive Manufacturing Techniques

### *7.2.1 Laser-Based Additive Manufacturing*

Powder bed fusion and directed energy deposition are the two commonly used AM techniques which uses lasers source to deposit the layer of the required material [7]. Laser additive manufacturing or LAM processes are divided into three categories SLA, SLM, and SLS.

Selective laser melting or SLM uses fiber lasers to melt the layers of the metal in the fine powder form to create a final layered 3D structure. Selective laser sintering or SLS uses carbon dioxide-based lasers to sinter the polymer-based material such as polyamides in the powder form and thus producing a layered 3D object. Stereolithography or SLA is a type of process in the vat photo-polymerization AM technique which uses a laser to harden the specific cross section on the surface of the resins placed in a vat or container to form solid layers of the 3D object [8].

All the three processes result in a smooth surface of the build object with high-resolution finishing. They can also be used to build large area object with high complexities. But, the only limitation with this process is that it requires expensive equipment with a support structure.

### 7.2.2 *Digital Light Processing*

In comparison to stereolithography, digital light processing (DLP) is a faster process. DLP cures the resin selectively in a pattern using a digital light projector to cover the entire layer at once, instead of tracing the surface. The digital projector results in the formation of square pixels of the required picture on each layer. Thus, each layer is made up of little rectangular bricks also called as voxels. DLP produces high-resolution objects and is flexible in using various types of resins for a specific application [9]. Just like stereolithography, DLP also requires post-processing steps and consists of expensive equipment.

### 7.2.3 *Two-Photon Polymerization*

Two-photon polymerization or 2PP is a technique used for micro-manufacturing 2D and 2.5D nano- and micro-scale structures via highly precise optical components. It utilizes nonlinear optical nature of light absorption to excite the photosensitive material, which undergoes a liquid to solid phase change via the polymerization process [10]. During the process, femtosecond pulsed lasers and numerical aperture lenses are utilized to enhance the probability of multiphoton absorption. Thus, the photosensitive material will absorb two or more photons to create the active species which further starts polymerization. With all these conditions, the area having highest light intensity will absorb multiphoton and confine the polymerization process within the volume of voxel or focused laser beam. It results in obtaining a 3D microstructure by precisely overlapping voxels via scanning the beam. It produces smooth and uniform shape with high accuracy in a single fabrication process. It also gives flexibility in choosing the design of the microstructure obtained.

## 7.3 Enhancing Precision and Resolution

Depending upon the specific applications, it is imperative to modify the elemental properties of the required material with optical solutions to print whether plastic parts or metal with high precision and resolution. Optical solutions involve using the light-based sources such as lasers, LEDs, which helps in evolving the AM processes efficiently.

Using spectrometers in the AM techniques can effectively ensure accurate and consistent analyses throughout the processes with inline measurements. Also, by integrating Optical Emission Spectroscopy (OES) with the AM technique can improve the quality control of the obtained 3D structure with deeper insights for refining and optimizing the printing process.

### *7.3.1 High-Resolution Printing with Laser Systems*

Laser-based printing techniques help in reducing the distortions and shrinkage in the manufactured 3D objects and thus enhancing the mechanical properties of the device. It also helps in defect mitigation in the targeted material. Lithography-based AM techniques are the most common methods to produce high-resolution 3D object using laser systems. SLA, DLP, 2PP, and VPP are the few common approaches in lithography AM techniques in which a liquid photopolymer located at specific location on a vat has been cured by light activated polymerization to produce a solid 3D object [11].

### *7.3.2 Photonics for Micro- and Nano-scale AM Techniques*

2PP is the most promising AM technique for producing microstructures using photonics. Manufacturing a 3D object at micro- and nano-scale requires highly precise printing nozzles to deposit nanoparticle solution or inks at the required surface on the target. Using a laser as a highly directional and focused light source can facilitate the formation of a required pattern. The most common application is the planar nano- and micro-level energy storage devices where inkjets printing helps in designing each layer with high precision and resolution [12].

## 7.4 Improving Material Properties

For an efficient AM technique, the manufacturer should design a process to optimize the technique parameters with respect to the properties of the targeted material to produce a uniform, stable and durable 3D object. To design or print a 3D object, it is important to improve the properties like physical, mechanical, etc., of the material with time to increase the lifetime of the printing object [13]. Various ways have been used over the decades to improve the material property using photonics, lets discuss a few of them below.

### 7.4.1 Optimizing the Material Curing via Photonics

Curing of a material using photonics can be employed in various sectors for coatings and printing a 3D object in milliseconds time. It is a method of using powerful light sources such as LASERS, LEDs, flash white light, in the infrared (IR) and ultraviolet (UV) spectra for drying inks (curing) and coatings [14]. For manufacturing nanoparticle-based thin films, photonic curing is one of the specialized methods as it uses intense light irradiation for rapidly curing and heating the material. Photonics curing is a three-step process: (i) formulating the paste or ink, (ii) drying the paste via evaporating the solvent, (iii) exposing powerful light irradiation on the targeted material to produce a dense film. It also helps in creating a gradient of the deposited material by heating the material in a short interval of time or by providing non-equilibrium heating.

Optimizing the curing process depends upon the chemical composition of the nanoparticles and their shape, distribution size, and degree of agglomeration. Changing anything in the process like time, pressure, temperature, etc., can cause a substantial variation in the targeted material in terms of specific surface area and grain boundaries. Thus, carefully optimization of the parameters for the process with respect to the material properties will result in the production of an efficient 3D object [15]. Weidling et al. have uses xenon flashlamp as the source for photonic curing to convert oxide sol-gels into an efficient semiconductors thin film for thin film transistors [16].

### 7.4.2 Effects of Light on Polymerization and Material Strength

As per the recent studies in 2019 at university of Michigan, researchers have reached a new approach of pattering using light source for polymerization inhibition volumes which was generated by irradiation at one wavelength will spatially confines the photopolymerized area via second concurrent irradiation wavelength. This process helps in controlling the hardness of the used resins and thus to modify it in a more complex 3D patterns in a single exposure. Irradiation of light sources helps in targeting a particular spot on the material with a high power, which maybe a defect, for healing and enhancing the material strength.

## 7.5 Speed and Efficiency Gains

### 7.5.1 *Accelerated Printing Processes Through Photonics*

In all the AM techniques, using photonics will accelerate the printing process. Using digital platform as the mold with an optically characterized CAD file will fully describe the structure of the obtained object. It actually reduces the number of steps involves in the fabrication process and results in producing highly precise, efficient, and innovative material.

### 7.5.2 *Reduction of Processing Times with Laser and Light-Based Techniques*

High power irradiation of a light source can effectively help in curing and bonding of a targeted material in a millisecond process. Light sources such as LASERS, flashlight, LEDs, etc., help in targeting a particular spot on the material via digital platform [17]. Any defect, stability, durability, uniformity, etc., of the material can be correct in a way lesser time as compared to the traditional manufacturing techniques.

## 7.6 Cost Reduction and Economic Impact

AM techniques with integrated photonics have the capability of generating 3D objects with an elaborative and complex architectures. These techniques greatly reduce the wastage of the targeted materials as compared to traditional AM techniques. Digital platform helps in reducing the physical stock level and enhances the working capital as per the demand in the market [18]. Also, fabrication of complex textures or optical shapes has become economical and easy by using a defined AM technique.

### 7.6.1 *Cost–Benefit Analysis of Photonics Integration*

Over a last decade, many photonic components like photonic crystal have been developed to enhance the optical resolution of the manufactured 3D objects. In making 3D objects like drones, robots, or any autonomous moving objects, integration of optical components in the techniques helps in compensating the aberrations for a clear vision with a real-time monitoring of the physiological parameters.

### 7.6.2 *Potential for Reducing Material Waste and Energy Consumption*

AM techniques integrated with photonics have efficiently reduced the waste production with lesser number of production steps during printing process as compared to conventional manufacturing techniques. With the advancement in the technology, electronic sector produces large amount of material wastes. Printing a layer by layer of a material have reduces the overall size of the device to nano- or micro-scale and thus minimizing the energy consumption during the process as well [15].

## 7.7 Challenges and Limitations

### 7.7.1 *Technical Challenges in Photonics-Based AM*

With all the benefits of producing complex, flexible, economical 3D structures, AM techniques face some challenges in the microfabrication processes. During this process, it is difficult to produce larger optical components. Also, writing in the high precision mode will come at the cost slow writing speed at the targeted layer. Another challenge is the scanning speed of the laser which determines the overall throughput in the process and thus the cost/part built.

In the 3D printing process, shape, size, and material used for the nozzle play a vital in enhancing the material's comprehensive strength and interlayer bonding. Any issue occurring in the nozzle may lead to deforming the printing material. Thus, designing a nozzle for a technique is the core of the technology [19].

### 7.7.2 *Addressing Material Compatibility and Process Stability*

Lasers-based AM techniques commonly face two types of challenges- improved quality part and the cost of the built-in parts. It is difficult to accurately steer the direction of the laser, scanner, and the controller selection at the target location during the process. A precise controlled over the three steps will justify the geometric part of the build object. Also, to ensure the formation of the 3D object with homogenous properties, the system requires a high-quality part having uniform laser density.

Roussel et al. have concluded that during the printing process, distance between the previous layer and nozzle may affect the flowability of the process which results in a deformed filament of the printing material [20]. Although, Shakor et al. had done experiment with different types of nozzles which leads to different intercepting shocks which enhances the shape and flowability of the slurry for the printing [21].

Limited type of the materials that can be used in the AM process and the poor precision caused due to "stair-stepping" effect may result in non-uniformity and instability of the manufactured 3D objects.

## 7.8 Recent Innovations and Future Directions

Photonics integrated AM techniques have palpable benefits. Along with CAD and new mechanical material technology, a complex architecture can be manufactured as an automated printing process. Yu et al. had showed that 3D printing objects shows shorter payback time period on investment, and with AI-based quality control, precision, and assurance, the process reduces the labor intensity [22].

### *7.8.1 Cutting-Edge Research in Photonics-Based AM Techniques*

Integration of photonics in a technique has expanded the region of working where miniaturization is required like in medical and space applications [23]. The optical components have the potential to generate high energy, directional pulses for depositing the material at a target and curing the defect as well. Thus, integration of photonics or merging optics with the AM techniques and electronic devices has revolutionize the whole sector [24].

A Chinese startup has developed a 3D metal printer named "precision-100," with a propriety feeding technology. It has ultra-fine metal powder bed and a precise laser to print with a correctness of 2–5 microns and eliminates the requirement of post-processing methods. Another Singaporean startup provides manufacture an on demand selective laser sintering to print nylon components with complex architecture and other functional parts with enhanced mechanical properties. They provide 0.2 mm dimensional accuracy with wall thickness of 1.0 mm. For a large-scale production of 3D objects, Printinue a German startup provides an automated printer for a continuous printing. It results in serial production of highly accurate and précised quality parts and products.

Recycling precious metals into a high-quality powder is another approach by an Italian startup "f3nice" to enhance the sustainability in metal printing 3D objects. It employs AM feedstock production in a circular approach. "Foundation Alloy," a US startup software enabled vertically integrated method to enhance flexibility and efficiency of supply chains in metal part production which helps in reducing the energy consumption. "Shape 1 + Dental" is one of the dental models based on DLP 3D printers which has been manufactured by "RAYSHAPE" a UK-based startup. They ensure high surface quality, precise layer thickness with accurate movement.

"Nematic 3D printing," developed by "NematX" a Swiss startup, employs a liquid crystal polymer (LCP) to provide a high inertness during polymer matrix reinforcement. This technique finds application in various sectors like electronics, medical and aerospace manufacturing. For the robotics industry, "Karkhana.io" an Indian startup has developed fused deposition modeling (FDM)-based 3D printing services. They provide a web interface to the manufactures for uploading the design parts and further choosing the required material. This technique supports various types of chemicals like acrylonitrile butadiene styrene, polylactic acid, polycarbonate, and polyethylene terephthalate glycol.

### *7.8.2 Emerging Trends and Future Prospects in Photonics-Enhanced AM Techniques*

Photonics integrated AM techniques have a very bright future prospect in the field of biomedical, automative, aerospace, and energy devices. AM processes have a capability to manufacture complex and unique 3D novel objects by tailoring material properties. However, the designers need to overcome the common challenges to take fully advantage of these capabilities. Following are some suggested developments to address the above challenges [25]:

(a) Empirical research should be done by the designers for optimizing the shape, size, and material used for the nozzle to improve the quality of the printing material.
(b) In the fields like game avatars, housewares, toys, collectables, etc., the technicians should design a CAD system for non-experts so that they can also take the advantage of the technology.
(c) There should be some standard models and methods to successfully design the 3D object despite of the property, shape, process, or other variabilities.
(d) To manufacture complex geometries with multiple materials, a new model of CAD systems is required to overcome the limitations of boundary representation and other parameters.
(e) Virtual aided designs are needed to assess the multifunctional design and impacts of parts and also the lifecycle cost of the produced objects.
(f) Real-time control system adaptive with closed loop algorithm is required for predictive models to enhance the stability of the manufactured device.
(g) Screening methodologies for selecting among the new materials available in the market for all the AM techniques.

## 7.9 Real-World Case Studies and Applications

Over a decade, most of the Asian countries have invested in cutting edge research facilities for photonics integrated AM techniques to foster robust collaboration between academics and industries. Scientist at Newport's technology and Application center has built the laser micro-FAB to produce highly precise and complex 3D micro-patterns. Another approach has been made by BMW; an automobile company has manufactured various parts in automobile like guide rails and convertible roof brackets via printing.

Su et al. highlight that using high-grade concrete as ink, theoretically a 3D light weight building can be made with a high quality to resist the earthquake. In this project, they utilize the recycle processed waste materials and natural stones collected from the mining sites [26]. In 2014, a five-story building in Jiangsu Provine, China, was built also known as the world's tallest 3D object printing. Another project presenting a two-story villa was built with factory manufactured LEGO bricks in three hours via mobile crane [27–29].

## 7.10 Conclusion: Recap of Key Insights

AM technology is a very economical and advance technology having potential applications in the field of energy devices, automative, aerospace, biomedical, etc. Integration of photonics with the AM technology has broaden the path for manufactures and designers to produce customized products with complex architecture in a time and process efficient way. This technology utilizes a digital platform or a CAD system to design a virtual prototype of the 3D object to analyze its every aspect like stability, durability, efficiency, precision, and resolution. Using the light sources such as LASERS, LED, flash lights may have help in targeting the specific area of the object to create the final product with the lesser wastage of the material used in the process. Photonic curing is one of the specialized cases of manufacturing a 3D object by coating and heating (high temperature) the required material on a specific target. It also helps in defect mitigation and correcting it.

Although, photonics integrated AM technology has numerous capabilities and advantages, it needs to overcome some of the limitations like requirement of specific types of material in a technique, costly, and bulky for a small-scale production. Achieving consistent outcomes while maintaining the quality of the product is the biggest challenge in the AM technology. In the architecture and construction industry, selecting the requisite material for the process is another tough task for a particular AM technique. To handle those challenges, there is a frequent need to collaborate industry and research field for tailoring the best solution for enhancing the efficiency of the printing technology.

## References


1. Abdulhameed O, Al-Ahmari A, Ameen W, Mian SH. Additive manufacturing: challenges, trends, and applications. Adv Mech Eng. 2019;11(2):1–27. https://doi.org/10.1177/1687814018822880.
2. Kanishka K, Acherjee B. Revolutionizing manufacturing: A comprehensive overview of additive manufacturing processes, materials, developments, and challenges. J Manuf Process. 2023;1(107):574–619. https://doi.org/10.1016/j.jmapro.2023.10.024.
3. Gao W, Zhang Y, Ramanujan D, Ramani K, Chen Y, Williams CB, et al. The status, challenges, and future of additive manufacturing in engineering. Comput Aided Des. 2015;69:65–89. https://doi.org/10.1016/j.cad.2015.04.001.
4. Gao H, An J, Chua CK, Bourell D, Kuo CN, Tan DTH. 3D printed optics and photonics: processes, materials and applications. Mater Today. 2023;1(69):107–32. https://doi.org/10.1016/j.mattod.2023.06.019.
5. Gong X, Anderson T, Chou K. Review on powder-based electron beam additive manufacturing technology. Manuf Rev. 2014;1:2. https://doi.org/10.1051/mfreview/2014001.
6. Somers P, Münchinger A, Maruo S, Moser C, Xu X, Wegener M. The physics of 3D printing with light. Nat Rev Phys. 2024;6(2):99–113. https://doi.org/10.1038/s42254-023-00671-3.
7. Camposeo A, Persano L, Farsari M, Pisignano D. Additive manufacturing: applications and directions in photonics and optoelectronics. Adv Opt Mater. 2018;7(1):1800419. https://doi.org/10.1002/adom.201800419.
8. Huang J, Qin Q, Wang J. A review of stereolithography: processes and systems. Processes. 2020;8(9):1138. https://doi.org/10.3390/pr8091138.
9. Huang P, Fu H, Tan MWM, Jiang Y, Lee PS. Digital light processing 3d-printed multilayer dielectric elastomer actuator for vibrotactile device. Adv Mater Technol. 2023;9(2):2301642. https://doi.org/10.1002/admt.202301642.
10. Bock S, Rades T, Rantanen J, Scherließ R. Additive manufacturing in respiratory sciences—current applications and future prospects. Adv Drug Deliv Rev. 2022;186: 114341. https://doi.org/10.1016/j.addr.2022.114341.
11. Melentiev R, Harakály G, Stögerer J, Mitteramskogler G, Wagih A, Lubnieau G, et al. High-resolution metal 3D printing via digital light processing. Addit Manuf. 2024;85(1):104156–6. https://doi.org/10.1016/j.addma.2024.104156.
12. Wang Z, Cheng Z, Zhang Y, Yu Y, Zhai X, Zhao Z, et al. Laser printing-based high-resolution metal patterns with customizable design and scalable fabrication of high-performance flexible planar micro energy storage devices. Chem Eng J. 2022;429: 132512. https://doi.org/10.1016/j.cej.2021.132512.
13. Bhuvanesh Kumar M, Sathiya P. Methods and materials for additive manufacturing: a critical review on advancements and challenges. Thin-Walled Struct. 2021;1(159):107228. https://doi.org/10.1016/j.tws.2020.107228.
14. Slimani MA, Cloutier SG, Izquierdo R. Recent advances in the photonic curing of the hole transport layer, the electron transport layer, and the perovskite layers to improve the performance of perovskite solar cells. Nanomaterials. 2024;14(10):886–96. https://doi.org/10.3390/nano14100886.
15. Altay BN, Turkani VS, Pekarovicova A, Fleming PD, Atashbar MZ, Bolduc M, et al. One-step photonic curing of screen-printed conductive Ni flake electrodes for use in flexible electronics. Sci Rep. 2021;9(11):3393. https://doi.org/10.1038/s41598-021-82961-3.
16. Weidling AM, Turkani VS, Luo B, Schroder KA, Swisher SL. Photonic curing of solution-processed oxide semiconductors with efficient gate absorbers and minimal substrate heating for high-performance thin-film transistors. ACS Omega. 2021;6(27):17323–17334. https://doi.org/10.1021/acsomega.1c01421.
17. Lee H, Lim CHJ, Low MJ, Tham N, Murukeshan VM, Kim YJ. Lasers in additive manufacturing: a review. Int J Precis Eng Manuf Green Technol. 2017;4(3):307–22. https://doi.org/10.1007/s40684-017-0037-7.

18. Calignano F, Mercurio V. An overview of the impact of additive manufacturing on supply chain, reshoring, and sustainability. Clean Logist Supply Chain. 2023;7: 100103. https://doi.org/10.1016/j.clscn.2023.100103.
19. Sepasgozar SME, Shi A, Yang L, Shirowzhan S, Edwards DJ. Additive manufacturing applications for Industry 4.0: a systematic critical review. Buildings. 2020;10(12):231. https://doi.org/10.3390/buildings10120231.
20. Roussel N, Spangenberg J, Wallevik J, Wolfs R. Numerical simulations of concrete processing: from standard formative casting to additive manufacturing. Cement Concrete Res. 2020;135: 106075. https://doi.org/10.1016/j.cemconres.2020.106075.
21. Shakor P, Nejadi S, Paul G. A study into the effect of different nozzles shapes and fibre-reinforcement in 3d printed mortar. Materials (Basel). 2019;12(10):1708. https://doi.org/10.3390/ma12101708.
22. Yu X, Yang J, Hu Z. Talking about the application of 3d printing building technology. Shanghai Build Mater. 2017;3:33–5.
23. Pérez M, Carou D, Rubio EM, Teti R. Current advances in additive manufacturing. Procedia CIRP. 2020;88:439–44. https://doi.org/10.1016/j.procir.2020.05.076.
24. Butt MA, Mateos X. Strategic insights into integrated photonics: core concepts, practical deployments, and future outlook. Appl Sci. 2024;14(14):6365. https://doi.org/10.3390/app14146365.
25. Guo N, Leu MC. Additive manufacturing: technology, applications and research needs. Front Mech Eng. 2013;8(3):215–243. https://doi.org/10.1007/s11465-013-0248-8.
26. Su Y, Li C, Yang T, Yang K, Li J. Study on application of 3d printing concrete technique in architectural engineering. Archit Technol. 2017;48:98–100.
27. Delgado Camacho D, Clayton P, O'Brien WJ, Seepersad C, Juenger M, Ferron R, et al. Applications of additive manufacturing in the construction industry—a forward-looking review. Autom Constr. 2018;89:110–9. https://doi.org/10.1016/j.autcon.2017.12.031.
28. Stott R. Chinese company constructs the world's tallest 3d printed building. 2015. Available online:https://www.archdaily.com/591331/chinese-company-creates-the-world-s-tallest-3d-printed-building.
29. Wang L. Chinese company 'builds' 3D-printed villa in less than 3 hours. 2015. Available online: https://inhabitat.com/chinese-company-builds-3d-printed-villa-in-less-than-3-hours.

# Chapter 8
# Green Photonics Enhancing Supply Chain Management

**Kavita Rani Segwal, Lalit Sharma, Arvind Sharma, and Sugandha Gupta**

**Abstract** 'Green Photonics' deals with a wide range of technologies like photovoltaic energy production, lighting and displays, communications, optical sensing applications for increased energy efficiency and reduction in environmental pollution, and laser processing of materials. The growth of the photonics market has been propelled by rising demand for photonics in various areas like aerospace, health care, and the environment. The use of laser technology for greenhouse gas monitoring, cost-effective cancer screening, water management, and growth of R&D activities related to photonics technology, which majorly deals with optical fibre communication, bio-photonics, and displays are also a driving force for the boom in the photonics supply chain. Integrating photonics with electronics and technological innovations has been a fascinating factor in transforming supply chain efficiency. Sustainable practices in photonics play a significant role in enhancing supply chain resilience. Through diversifying sources and minimizing dependency on non-renewable resources, companies can better identify and manage risk of supply chain disruptions and improve supply chain management. Funding and investment in photonic technology projects boost the photonics market. The prominent influence of the green photonics industry has led to organizations such as Optical Society of America (OSA) and Opto-Electronics Industry Development Association (OIDA) organizing seminars and representing special reports on the contribution of Green Photonics in the optoelectronic market. According to a special report of OIDA, the share of green Photonics is very encouraging ~ 19%, and by 2025, 3D sensing technologies, LiDAR, and AR/VR displays will increase by much higher compounded annual growth ratios (CAGRs). According to market predictions, by 2032, the global Photonics market will grow from $984 billion to $1643 billion at a CAGR of ~ 7%. These predictions represent green photonics as a driver for growth and employment, opening the scope of international collaborations among global leaders to take a quantum leap in the photonic industry.

K. R. Segwal (✉) · A. Sharma · S. Gupta
Bhagwan Parshuram Institute of Technology, Sector-17, Delhi, India
e-mail: kavita.segwal@gmail.com

L. Sharma
Concentrix Technologies Services, Oshawa, ON, Canada

 
V. Bhardwaj et al. (eds.), *Photonics and Optoelectronics in Industry 5.0*, Advances in Optics and Optoelectronics, https://doi.org/10.1007/978-981-96-7594-4_8

**Keywords** Green photonics · Sustainability · Optoelectronics · Photonics market · Supply chain dynamics

## 8.1 Introduction

'Photonics' refers to the science and technology of producing, controlling, and detecting light waves and photons. It refers to the technology of producing and harnessing light energy. It covers a broad electromagnetic wavelength spectrum from gamma rays to infrared light. The applications lie in diverse areas, including universe exploration, diagnosis and disease treatment, and security assurance through smart devices. The term 'photonics' comes into play when photon energy exceeds the thermal energy of material particles. In electronics, Opto-microelectronic techniques are used for storage and processing, while laser technologies are gaining popularity due to precision in manufacturing and metrology [1, 2]. 'Green photonics' refers to an energy-efficient solution that tends to reduce greenhouse gas emissions or yield environmentally sustainable solutions [3] and improves health care. An example of 'Green photonics' is a photovoltaic solar cell that uses sunlight to produce clean electricity. This includes solid-state lighting (SSL), photovoltaic solar cells, data communication, optical sensing, and laser processing of materials. But now, the lighting sector is shifting to new digital technologies [4] developed around light-emitting diodes (LEDs) and organic LEDs (OLEDs), and the most significant impact on energy efficiency and sustainability. This field significantly contributes to sustainability by delivering improved economic, societal, and environmental outcomes in industrial product design and production technologies [1]. Green photonics is an emerging field that focuses on the development and application of photonic technologies to promote environmental sustainability. This includes innovations in energy-efficient lighting, solar energy harvesting, and optical communication systems that reduce power consumption. By leveraging advanced materials and designs, green photonics aims to minimize the ecological footprint of photonic devices and systems (Table 8.1).

### *8.1.1 Emerging Applications of Photonics*

Recent technologies [4] such as autonomous vehicles, 3D sensing, and optical interconnect, are integral to Photonics. Photons are analogous to electrons in electronics and provide 'vision' for futuristic machines. Key parts include Si and silicon nitride (SiN) waveguides, MEMS devices, GaAs & GaN LEDs, optical modulators, InP and GaAs lasers, and Si and indium gallium arsenide (InGaAs) photodetectors. GaAs lasers are essential for 3D sensing applications, while the demand for micro-LED applications is increasing, promptly improving wafer yields and process control. However, research on thin-film technologies with high optoelectric coefficients aims

**Table 8.1** Photonics technologies and their impact [2]

| Technology | Devices | Applications | Consequences |
|---|---|---|---|
| Photovoltaics | CdTe, aSi, pSi, | Power production | Reduced greenhouse gas emissions and renewable energy |
| Solid state lighting | LEDs and OLEDs, | Light displays | Energy conservation |
| UV disinfection | UV LEDs | Water purification | Improved quality of water and pollution |
| Optical transceivers | VCSELs, PIN detectors | Digital data centres | Minimal energy usage |
| Optical sensors | Optical fibres, detectors, Bragg reflectors | Harnessing energy, environmental monitoring, and gas sensing | Reduced pollution and carbon emission |
| Power displays | LCDs, OLEDs, LEDs, and MEMs | Light displays | Reduced energy consumption |
| Green optoelectronic manufacturing | LEDs, photodiodes, photodetectors, photoresistors | Light-sensitive switching, harnessing solar energy | Low energy consumption, reduced greenhouse gases and pollution |

to increase optical interconnects' speed and bandwidth envelopes. Photonics technologies are revolutionizing medical diagnostics and treatment. Techniques like optical coherence tomography (OCT) and multiphoton microscopy provide high-resolution imaging for early disease detection and monitoring. Photonics plays a crucial role in the development of quantum computers. Photonic qubits, which use photons to represent quantum information, offer advantages in terms of speed and scalability. This technology has the potential to revolutionize computing by solving complex problems much faster than classical computers. LIDAR navigation and safety devices are used in advanced driver-assistance systems (ADAS) and autonomous vehicles in self-driven cars. End products are integrating lasers with sensors and optics, opening new opportunities for photonics manufacturers to foster supply chain management. The emerging technologies in the area of photonics are shown in Fig. 8.1.

(i) **3D Sensing**

Wide usage of smartphones retains more personal information; hence, stringent security beyond biometric authentication is required. After the launch of Apple's 2017 iPhone, vertical cavity surface emitting lasers (VCSELs) have gained significant popularity in the consumer market. Ten thousand laser beams shine on the user's face through the VCSELs. They are collected to create depth mapping of the face to produce a unique identifier for the person [5]. This technology has been extended with depth information using a time-of-flight sensor [6] utilizing VCSELs to generate a 3D space image. For example, a piece of artwork can now be virtually placed in a space to visualize it before buying it.

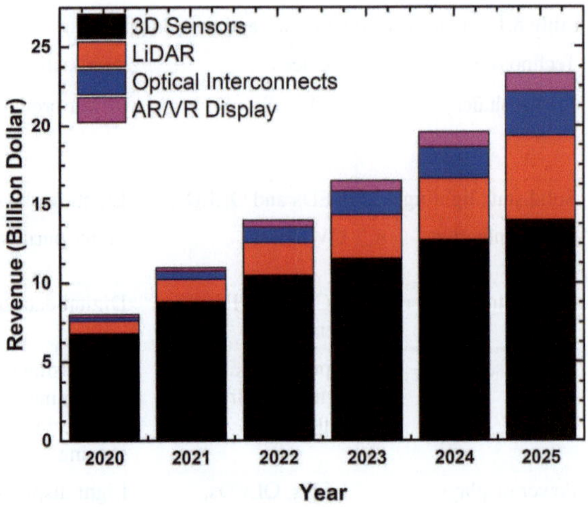


**Fig. 8.1** Growth trajectory of emerging photonic applications (*Source* Yole Développement reports)

(ii) **Optical Communication**

Data transport using electronic switches will not be able to fulfil future bandwidth and energy requirements due to the rise in global data traffic. Si photonics and InP-based optical interconnects have addressed these data transfer challenges. Si photonics uses semiconductor lasers and the effective bulk Si manufacturing process to provide faster data transport over large distances than traditional methods.

(iii) **LiDAR**

Due to its long-range, high-resolution, and 3D imaging capabilities, light detection and ranging (LiDAR) is an excellent solution for autonomous driving [4]. A well-established system for lasers and photodetectors in the 905 nm range is the most widely accepted of the two LiDAR frequency ranges, along with 1550 nm. While MEMS-based technologies have been included in the sophisticated driver-assistance systems of newer automobile models, beam-steering technologies such as mechanical rotation exhibit serious reliability issues. Chips with heterogeneous laser and detector integration must be manufactured in large quantities to meet the LiDAR system's performance and cost requirements. MEMS-based LiDAR technologies are now making a significant contribution to industry demands.

(iv) **Micro-LEDs**

Applications for augmented reality and virtual reality (AR/VR) products that require self-emissive red–green–blue (RGB) displays may be made possible using micro-LEDs. Achieving the necessary quantum efficiencies, transferring mass onto the substrate at a reasonable cost, and testing are some hurdles in RGB micro-LEDs. New backplane architectures, die transfer techniques, epitaxial growth optimization, substrate engineering, and creative device designs are being researched and developed to make micro-LED technologies competitive with the established liquid crystal display (LCD) and organic LED (OLED) technologies [7].

(v) **OLED**

OLED thin-film transistor (TFT) technology has advantages over liquid crystal displays (LCDs) because of self-illuminating OLED TFT pixels. OLED architecture leads to flexible and thinner panels with a wide colour range and viewing angles, negligible motion blur, higher contrast ratio, bright pixels, lower working temperatures, and reduced energy consumption [6]. OLED technology facilitates 'capture and display' ability, enabling users to highlight some specific region within an overall picture and scan a widescreen image at 2 K or 4 K resolutions. However, 8 K works with both OLED and LCD technology. Choosing OLED over LCD technology has been challenging in achieving price parity, higher yields, and panel manufacturing costs.

(vi) **Photonics Quantum Computing**

In past decade, quantum photonics has advanced rapidly, dealing with basic science and technology where quantum effects play a crucial role. Quantum photonics mainly focuses on quantized light-matter interaction which involves active control of light at quantum level. Quantum technologies have the potential to play significant roles in various applications such as quantum thermometry, quantum metrology, quantum information processing, and quantum bio-marking and sensing. Quantum computers are being extensively explored with progress in superconductors, ion traps, and silicon photonics circuitries where integration of quantum photonics plays a key role in realizing real quantum computers. Realization of on-chip quantum photonics is possible by Integrating quantum Photonics (IQP) with conventional CMOS fabrication processes to make quantum computing a reality [8].

### *8.1.2 Supply Chain Overview*

Supply chain management (SCM) [9] refers to monitoring and enhancing the production and distribution of goods and services. It efficiently improves the processes involved in converting raw materials and components into final products and launching them in the market. An efficient SCM helps streamline the process to eliminate waste and maximize competency and customer value in the market [3].

(a) **Supply Chain Management**

A robust supply chain incorporating environmental, social, and E-governance (ESG) criteria is greatly needed [9]. As per the customer's point of view, the expectation is to provide them with the parts and supplies they need, even during challenging and inconsistent conditions.

The environmental point of view describes it as repairing, recycling, and reusing parts instead of using scarce resources to produce new ones with less energy usage. Hence minimizing greenhouse gases and waste generated throughout the supply chain.

The corporate perspective enables a growth process for customers, investors, and the government to incorporate sustainability practices.

In the last decade, photonic applications have shown tremendous growth. The key applications such as optical communication, LiDAR, 3D sensing, and AR/VR displays have grown collectively from an $8B market. AR/VR displays are the products benefiting from micro-LED technology to $23.3B by 2025 at a CAGR of 31% (See Fig. 8.1) [4]. While 3D sensing technologies and LiDAR are expected to grow at much higher CAGRs. Addressing photonics device technology challenges in fabrication, system integration, and performance will boost the industry's growth.

**Top Players in the Photonics Supply Chain** [10] **are:**

- IBM
- AIO Core Co. Ltd.
- Cisco Systems, Inc.
- Intel
- Infinera Corporation
- Mellanox Technologies Ltd.
- IPG Photonics Corporation
- NKT photonics
- STMicroelectronics HV
- Hamamatsu Photonics K. K.
- Coherent
- Thorlabs
- Siemens
- Honeywell

(i) **Manufacturers**

- **Component Manufacturers**: These companies produce essential photonics components such as lasers, sensors, optics, and fibre optics. Notable players include **Coherent**, **Thorlabs**, and **Hamamatsu Photonics**.
- **System Integrators**: They assemble photonics components into complete systems for telecommunications, medical devices, and industrial automation applications. Companies like **Siemens** and **Honeywell** are key players in this domain.

(ii) **Suppliers**

- **Raw Material Suppliers**: The suppliers provide the necessary materials for photonics components, such as semiconductors and optical materials. Companies like **Corning** and **Schott AG** are the main suppliers [11].
- **Specialized Equipment Suppliers**: They offer machinery and tools for manufacturing photonics components, including laser cutting and optical testing equipment.

(iii) **Distributors**

- **Wholesale Distributors**: Photonics components and systems are distributed to various Companies through distributors like **Digi-Key** and **Mouser Electronics**, which make these products accessible to manufacturers and end users.
- **Value-Added Resellers (VARs)**: Resellers provide customization and integration services of photonics systems for specific applications.

(b) **Photonics Industry Versus Traditional Supply Chain**

(i) **Nature of Products**

- **Traditional Supply Chains**: Typically involve physical goods like consumer products, machinery, and raw materials. The focus is on tangible items that can be easily stored and transported.
- **Photonics Supply Chains**: It deals with highly specialized components such as lasers, optical fibres, and sensors. These products often require precise manufacturing processes and are sensitive to environmental conditions.

(ii) **Complexity and Integration**

- **Traditional Supply Chains**: It can be reactive, responding to demand based on historical data, and often operates linearly, with clear stages from raw materials to finished products.
- **Photonics Supply Chains** Are more integrated and complex, involving collaboration between manufacturers, suppliers, and research institutions. They often require real-time data sharing and adaptability to rapidly changing technologies and market needs.

(iii) **Supply Chain Resilience**

- **Traditional Supply Chains**: They often have established protocols for managing the risks of natural disasters or geopolitical issues.
- **Photonics Supply Chains**: It has been vulnerable to disruptions, especially in sourcing high-quality materials and components. Recent surveys indicate that over 80% of photonics companies face significant supply chain issues, highlighting the need for improved resilience [11].

(iv) **Technological Dependency**

- **Traditional Supply Chains**: Rely on conventional logistics and inventory management systems, which may not be as technologically advanced.
- **Photonics Supply Chains**: Heavily depend on cutting-edge technologies, including integrated photonics and advanced manufacturing techniques. Technological reliance can lead to faster innovation cycles but requires consistent adaptation [11, 12].

(v) **Market Dynamics**

- **Traditional Supply Chains**: Generally operate within established markets with predictable demand patterns.

- **Photonics Supply Chains**: These are influenced by rapidly evolving sectors like telecommunications, health care, and consumer electronics, leading to fluctuating demand and the need for agile responses [13].

(c) **Challenges in the Supply Chain**

The photonics supply chain faces several challenges, including:

- **Supply Chain Disruptions**: Raw material shortages and delivery delays are major issues in the supply chain. Additionally, buying concerns such as material deficiency, manufacturing capabilities and machineries, poor retention of skilled employees, shipping and warehouse management constraints, and customer's political agitations are dominant challenges in photonics supply chain [14].
- **Technological Advancements**: Rapid advancements require constant adaptation and innovation within the supply chain to meet evolving market demands.
- **Geopolitical Turmoil**: Geopolitical issues have a wide impact on photonics industry including trade and regional conflicts and instabilities. These factors affect the availability and cost of essential goods leading to long term movements in photonics supply chains.
- **Global Dependency**: Any disruption in complex global supply chain can have a disturbing effect on photonics industry, causing delays and shortages in many parts of the world. Due to this dependency, the industry becomes vulnerable to international trade policies and economic conditions.
- **Environmental regulations and Sustainable practices**: Increasing demand to adopt sustainable practices and comply with environmental regulations enhances complexity in supply chain. These demands need to be balanced through maintaining efficient and cost-effective supply chains.
- **Workforce and Skill Gaps**: Deficiency of skilled workers particularly in areas of advanced manufacturing and supply chain management can hinder the innovations and progress in Photonics industry.

## 8.2 Integration of Photonics and Electronics

An alternative substitute for charged carriers has been provided by Photonic integrated circuits (PICs) for signal transmission using light [13]. For processing and transmitting data over long distances with minimal losses, higher speed, and low energy consumption, various components such as optical fibres, waveguides, and photonic crystals are used in PICs [15, 16].

### 8.2.1 Challenges in Integrating Photonic Components with Electronic Systems

Various bottleneck challenges must be addressed in developing PICs for the full potential usage of PICs, including thermal management, material compatibility, scalability, and fabrication precision. Some of them are discussed below.

- The foremost challenge is material compatibility [17], a crucial aspect of fabricating efficient and reliable photonic devices, including a high refractive index. Combining silicon and III–V semiconductors (e.g. indium phosphide and gallium arsenide) can improve emission of light [18]. This is because silicon excels in electronic properties but lacks light emission because of indirect bandgap. Also, material compatibility can be overcome by using new materials such as polymers and $Si_3N_4$ (Silicon Nitride) [19].
- Precision in fabricating photonic device structures [20], such as resonators and waveguides, is another challenge for achieving sub-wavelength accuracy for proper light confinement and propagation [21]. This led to the need for advanced lithography, etching techniques, and metrology tools for higher resolution. It mainly focuses on process control that minimizes fabrication deviation and precision [22] with the help of AI and ML integration.

Precise temperature control becomes a critical challenge for thermal management for PICs operating at higher power [23]. Effective strategies are required to maintain device reliability and performance. These strategies include thermal sensitivity minimization [24] for designing photonic circuits and utilizing thermoelectric coolers and heat sinks to dissipate heat. Another challenge is achieving seamless photonic integration of electronic components and photonic devices while maintaining scalability. This has increased the demand for multifunctional PICs, making single chip integration of complex photonic devices difficult. There has been a need for modern design equipments capable of handling complex manufacturing thereby achieving higher production yields. The photonic industry has streamlined the process by using standardized design and automation platforms to streamline the process. An investment in research and development is as essential as adopting advanced simulation and modelling tools to predict the accurate behaviour of photonic devices. Integration of AI and ML in the design and manufacturing processes can further enhance efficiency and performance, enabling the rapid development of next-generation photonic devices [25].

Additionally, establishing industry standards and fostering collaboration across the photonics ecosystem will promote the best innovation practices and technological advancements. Through innovation and collaboration, the development of integrated photonic devices can help unlocking new areas and opportunities in telecommunications, sensing, and computing, etc. Continuous research and advancements in materials, design, and manufacturing techniques can continue to push the boundaries of integrated photonics, promising a future where photonics and electronics seamlessly work together.

### *8.2.2 Case Studies*

Integrating photonics and electronics is a fascinating field. Some real life studies are mentioned below:

(i) Convergence of Circuits: Researchers have been working on integrating photonics and electronics to overcome limitations in conventional electronic circuits. More efficient and high-performance systems have been fabricated by combining photonic components like waveguides and modulators with electronic circuits.
(ii) Electronic-Photonic Integrated Circuits (EPIC) Hardware Accelerator: A framework of hybrid design EPIC hardware accelerator has been proposed which uses architecture-level integrations of both photonic and electronic components to boost the performance of device.
(iii) 3D Integrated Circuits: 3D integrated circuits are able to resolve critical input/output limitations in computing chips. This enables the production of larger and more complex chips essential for futuristic data transmission and high-performance systems.
(iv) Mass Production of Optoelectronic Systems: Mass production of optoelectronic systems on chips revolutionized the fabrication of photonic devices on a bulk silicon substrate alongside millions of electronic transistors.

These case studies highlight the potential of integrating photonics and electronics to create more efficient, scalable, high-performance systems.

## 8.3 Technological Innovations enhancing Supply Chain Efficiency

Technological innovations have significant impact on efficiency of the supply chain. Some of the key impacts are:

(i) **Improvement in Visibility and Transparency**: Advanced digital technologies have improved inventory and demand visibility which allows companies in real-time tracking of products, reducing delays and better decision-making.
(ii) **Better Communication and Collaboration**: Cloud computing and IoT enable better communication and collaboration among supply chain partners, leading to more synchronized operations and efficient resource management.
(iii) **Automation and Robotics**: Automation technologies, including robotics and AI, in warehouse and logistics, help increase the productivity by reducing human errors and streamlining the repetitive tasks.
(iv) **AI Predictive Analytics**: AI predictive analytics help in accurate prediction about demand, optimizing inventories and reducing waste. This enables proactive management of supply chain disruptions.

(v) **Blockchain Technology**: Blockchain enhances the security and transparency of transactions within the supply chain, ensuring data integrity and traceability, such as in food and pharma industries.
(vi) **Resilience and Flexibility**: Supply chains have adopted more resilient and flexible inventory management, supported by advanced technologies for better handling volatility and disruptions.

These innovations collectively contribute to more efficient, responsive, and resilient supply chains, ultimately enhancing overall performance and competitiveness.

## 8.4 Photonics Supply Chain and Sustainability

Sustainable practices are crucial in the photonics supply chain because of certain reasons:

i. **Environmental Impact**: The photonics industry has a significant ecological impression, sustainable practices help reduce greenhouse gas emissions and wastes and conserve natural resources.
ii. **Laws and Regulations**: Governments and international bodies are enforcing strict Laws and environmental regulations for adopting sustainable practices to ensure stringent regulatory compliance, avoiding fines and legalities.
iii. **Cost Reduction**: Sustainable supply chain practices results in long term cost efficiency. It helps lowering the operational cost through optimizing energy use, reducing waste, and improving resource efficiency.
iv. **Enhancing Reputation and Brand Value**: Companies prioritizing sustainability have a better reputation and increased brand loyalty. Consumers and stakeholders are more likely to support businesses committed towards environmental and social responsibility.
v. **Risk Management**: Sustainable practices can enhance the durability of the supply chain, by diversifying sources and reducing dependency on non-renewable resources, companies can better manage risks associated with supply chain disruptions.
vi. **Innovation and Competitiveness**: Encompassing sustainability drives innovation, new technologies and processes that are more efficient and environmentally friendly, giving companies a competitive advantage in the market.

## 8.5 Future Trends in Photonics and Supply Chain Management

The future of photonics is set to be revolutionary, driven by advancements in integrated photonics, quantum technologies, and artificial intelligence. Because of increasing demand for high-speed data transmission and advanced manufacturing, usage of photonic components will become significant. Expanding semiconductor boundaries will help meet the rising demands of photonic components by enhancing the production capabilities. The Photonics supply chain is projected to become more resilient and localized, reducing reliability on a few key suppliers, mitigating risks from geopolitical and global turmoils. Sustainability will also play a crucial role in developing energy-efficient photonic devices. Emerging areas of research like topological photonics and AI-empowered nanophotonics will continue to drive innovation, leading to new applications and breakthroughs [26].

Integrated photonics, which involves integrating photonic components on a single chip, is anticipated to significantly enhance performance and reduce costs, making photonic systems more accessible for various applications. The construction of new semiconductor fabs, driven by initiatives like the US CHIPS and Science Act, will further bolster the production capacity for photonic components. Moreover, the focus on green photonics will lead to the development of sustainable practices within the industry, contributing to energy-efficient solutions. Research in non-Hermitian photonics and AI-empowered nanophotonics is expected to drive further innovation, opening up new possibilities and applications [27, 28].

### *8.5.1 Market Outlook*

The global photonics market is projected to grow significantly, driven by increasing demand in sectors like telecommunications, health care, and manufacturing. By 2025, the market for photonics-enabled systems is expected to approach \$2 trillion [29]. In 2023, Asia Pacific dominated the photonics global market with a market share of 63.2% (See Fig. 8.2). The global Photonics market is projected to boom from USD 983.51 billion in 2024 to USD 1642.61 billion by 2032 at a Compounded Annual Growth Rate (CAGR) of 6.7% [30]. According to a special report of OIDA, the Green Photonics share is very encouraging ~ 19.6%, while the CAGR for global optoelectronics is 3.4%. This resulted in revenue of \$493 billion for the optoelectronics market by 2020, of which the green photonics market share was 53% [30]. The Asia Pacific photonics market is expected to witness a rapid growth as a result of the rising share of innovations and R&D. China has been at the forefront of advancements and innovations in photonics. Most leading Chinese companies are strong producers and exporters of photonics components having a strong supply chain worldwide. Rising demand of faster and efficient next-generation integrated devices is scheduled to propel market growth.

Fig. 8.2 Global photonics market share, 2023 (Data taken from fortunebusinessinsights.com)

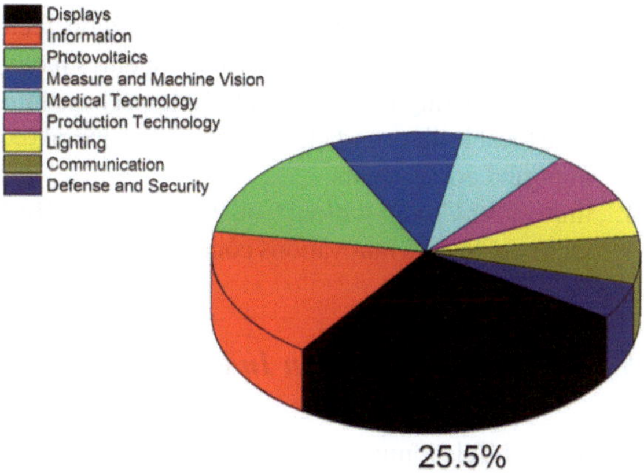


North American market is driven by tech giants, such as Facebook and Microsoft, requiring optimization of data processing and transmission. Photonics considerably improves the high-speed data processing. Additionally, photon-based quantum computers can operate at ambient temperature, which is advantageous over electron-based quantum computers [31].

## 8.6 India's Semiconductors Supply Chain Management

The Government of India has been investing heavily in semiconductor manufacturing, aiming to make it a global hub in the photonics industry. Several initiatives have been taken by the National Mission on Electronics and Semiconductor Manufacturing Corporation of India [32] to develop the semiconductor ecosystem by analysing semiconductor wafer design, manufacture, and final testing. Some bottleneck issues in the semiconductor industry are due to a lack of high capital costs, a skilled workforce, and a lack of government support through tax and subsidies, which make it costlier for small and medium-sized businesses. Despite these challenges, through significant investments and regulations, India could emerge as a key player in the global semiconductor market [31].

### *8.6.1 Distributor Partners' Role*

With the expansion of India's semiconductor industry, distributor partners' roles have become crucial in ensuring the efficient delivery and maintenance of semiconductor components. Both play key roles such as:

1. Supply Chain Management: The distributor's complex supply chain of semiconductors is streamlined to minimize disruption and ensure timely delivery.

2. Technical Support: Manufacturers provide technical support in prototyping, design, and engineering to local companies having less expertise.
3. Interconnecting Local–Global Market: These interconnect global manufacturers with local Indian markets to ensure that general requirements are fulfilled through imported components.
4. Innovative Collaboration: Collaboration with R&D centres, start-ups, and manufacturers fosters innovations contributing to the semiconductor hub's emergence.

### 8.6.2 *Recent Trends in India's Semiconductor Market*

1. 5G and Telecommunications: The demand for high-performance semiconductors for applications like 5G network deployment has become a key focus for companies like Bharti Airtel and Reliance Jio, driving significant investment in this sector.
2. Automotive and EVs: India's electric vehicle sector is a major consumer of semiconductor chips for controllers, sensors, and power systems.
3. IoT: The advancement of IoT in health care, agriculture, and smart cities is driving the demand for specialized semiconductor technologies.
4. AI and ML: There is a great need for semiconductors designed for AI/ ML technologies with the recent advancement in artificial Intelligence and machine learning.
5. Local Producers: The Government's assistance in semiconductor design can lead to establishment of local factories, enhancing domestic production.

## 8.7 India's Photonics Industry: A Comparison with Global Leaders

Compared to established countries like Taiwan, China, and South Korea, the Indian semiconductor sector is still in the initial stage. These established countries have made significant investments and emerged as global power hubs. Taiwan dominated the worldwide market with 60%, followed by South Korea, which had 20%. Meanwhile, China, the largest semiconductor consumer, is actively working to enhance and expand its semiconductor photonic industry. India's global semiconductor share is currently below 1%. However, India-like countries have a vast pool of talented scientists and engineers and relatively low business costs. The government of India has introduced many initiatives to attract investment in this sector, such as a production-linked scheme [31]. Many Indian companies are involved in semiconductor manufacturing, such as LED displays, telecommunication systems, power devices, solar cells and modules, automotive electronics, medical devices, and industrial automation devices.

Some Industries are:

**SPEL Semiconductors**: India's only company that provides IC assembly and test facilities is SPEL Semiconductors. Its services include wire bonding, wafer bumping, testing, and packaging.
**MOsChip Technologies Ltd**: This company provides cutting-edge designs and markets a variety of analogue and mixed ICs for automotive electronics, power management, and automation.
**e info chips pvt ltd**: Chip design, verification, and testing are some of the semiconductor design services offered by this company. It also provides a range of intellectual property (IP) blocks to expedite the design process.
**Tata Elxsi**: System-on-chip design, embedded software development, test and validation are some of the company's semiconductor engineering services.
**HCL Technologies**: The services include IT solutions for design and manufacturing.

India's supply chain can be strengthened through international collaborations and partnerships. India's prime partners include:

- United States: US companies can support India's Photonics market through chip design, production, and R&D collaboration.
- Japan: India can collaborate with another leading semiconductor producer for manufacturing, chip designing, and equipment supply.
- South Korea: South Korea's advanced chip manufacturing plants can help India with chip production and technology transfer.
- Taiwan: The leading global semiconductor producers, Taiwan and India, can collaborate on research, design, and chip manufacturing.
- Germany: Germany is leading in semiconductor materials and equipments. India can join hands with Germany to enhance its fabrication capabilities.

### 8.7.1 Global Key Developments

July 2023–the photonic industry has raised 100 million USD with the help of an AI platform. This funding will accelerate the R&D sector for further development and commercialization of products.

October 2023–Montana has many companies related to photonics, optics, and quantum computing. One of the brands, Gianforte, emphasized the high impact of the photonic industry, which is expected to promote global economic growth. Moreover, a partnership with Taiwan could further boost new opportunities.

March 2024: The Hitachi group and VLC Photonics partnered to provide design and testing services for photonic integrated circuits (PICs) utilizing the Open Light process design kit (PDK). This collaboration has facilitated progress and innovation, driving growth across various applications and the silicon photonics industry's market sectors [33].

## 8.8 Summary

Photonics significantly enhances supply chain management by improving various aspects such as data transmission, sensing, and automation. Using fibre optic cables, photonics ensure faster and more reliable data transfer, essential for real-time tracking and monitoring of goods. Advanced photonic sensors and imaging systems provide detailed information about the condition and location of products, aiding in quality control and reducing losses. Additionally, photonics supports the development of automated systems and robotics in warehouses and manufacturing plants, increasing efficiency, and minimizing human error. By integrating photonic technologies, companies can better predict and respond to supply chain disruptions, manage inventory levels, and optimize logistics, ultimately leading to more sustainable and resilient supply chain practices. Photonics contributes to energy-efficient operations, reducing the carbon footprint of supply chain activities. Leveraging photonics for precise monitoring and control also helps minimize waste and ensure that resources are used optimally. This integration of photonic technologies enhances operational efficiency and aligns with the growing emphasis on sustainability in supply chain management. The global photonics market is projected to grow due to its increasing telecommunications, health care, and manufacturing demand. By 2025, the market for photonics-enabled systems is expected to grow by **$2 trillion** [30]. The Indian government is also taking the initiative to bring its photonics market into the photonic industries race. Collaboration with other countries and R&D activities will boost the Indian photonics industry. In all, green photonics represents a significant step towards a more sustainable and energy-efficient future which will enhance supply chain management and photonics industry.

## References

1. Wessler B, Tober U. Green photonics: the role of photonics in sustainable product design. In: SPIE eco-photonics 2011: sustainable design, manufacturing, and engineering workforce education for a green future, 2011. p. 806515. https://doi.org/10.1117/12.882979.
2. Vijaylakshmi K, Chandraprabha C. Green photonics, a broad overview and future prospects. J Manage Eng Inf Technol (JMEIT). 2014;1:1–14.
3. David I. OIDA releases new green photonics market data (2009). www.oida.org/news/oidanews/2009/377 Feb 27 2009.
4. Les CB. Environmental Sensing market upwardly mobile, Photonics Spectra, 24 (Environmental sensing market upwardly mobile I light speed I Aug 2009 I Photonics Spectra) (2009).
5. Rai S. Photonic device technologies for emerging markets (appliedmaterials.com) (2020).
6. Kumar P. The ultimate guide to depth perception and 3d imaging technologies (2025). https://www.e-consystems.com/blog/camera/technology/the-ultimate-guide-to-depth-perception-and-3d-imaging-technologies/.
7. Dash S. MicroLED: Emerging as a next generation display technology—display daily. 2019.
8. Chen Z, Segev M. Highlighting photonics: looking into the next decade. eLight. 2021;1:2. https://doi.org/10.1186/s43593-021-00002-y.

9. Lele V, Ajmeri Y. From source to shelf. Springer Nature; 2024. https://doi.org/10.1007/979-8-8688-0986-6.
10. Global photonics market size, share & trends analysis report by product, application, regional outlook and forecast, 2024–2031 (2024), report id: KBV-26336, Photonics market size, share & top key players | 2031.
11. Introduction to supply chain management (2024). https://www.geeksforgeeks.org/introduction-to-supply-chain-management
12. Hausken T. Optica industry reports, optica market update: the outlook for supply chains, pricing, profits, and the market in 2022, optica industry report, 8. Optica market update: the outlook for supply chains, pricing, profits, and the market in 2022 (2022)
13. Bhutt MA, Mateos X. Strategic insights into integrated photonics: core concepts, practical deployments, and future outlook. Appl Sci. 2024;14(14):6365. https://doi.org/10.3390/app14146365.
14. Photonics industry supply chain survey 2023 (2023). Photonics21/EPIC.
15. Wang C, Li Z, Riemensberger J, Lihachev G, Churaev M, Kao W, Ji X, Zhang J, Blesin T, Davydova A, et al. Lithium tantalate photonic integrated circuits for volume manufacturing. Nature. 2024;629:784–90. https://doi.org/10.1038/s41586-024-07369-1.
16. Butt MA, Piramidowicz R. Orthogonal mode couplers for plasmonic chips based on metal–insulator–metal waveguide for temperature sensing application. Sci Rep. 2024;14:3474. https://doi.org/10.1038/s41598-024-54244-0.
17. Homes CC, Carr GL, Lobo RPSM, LaVigne JD, Tanner DB. Silicon beam splitter for far-infrared and terahertz spectroscopy. Appl Opt. 2007;46:7884–8. https://doi.org/10.1364/ao.46.007884.
18. Butt MA, Khonina SN, Kazanskiy NL. 2D-Photonic crystal heterostructures for the realization of compact photonic devices. Photonics Nanostruct Fundam Appl. 2021;44:100903. https://doi.org/10.1016/j.photonics.2021.100903.
19. Bogdanov S, Shalaginov MY, Boltasseva A, Shalaev VM. Material platforms for integrated quantum photonics. Optical Mat Express. 2017;7(1):111–3. https://doi.org/10.1364/OME.7.000111.
20. Klamkin J, Zhao H, Song B, Liu Y, Isaac B, Pinna S, Sang F, Coldren L. Indium phosphide photonic integrated circuits: technology and applications. In: Proceedings of the 2018 IEEE BiCMOS and compound semiconductor integrated circuits and technology symposium (BCICTS), San Diego, CA, USA, 15–17. 2018. pp. 8–13. https://doi.org/10.1109/BCICTS.2018.8550947.
21. Liang D, Roelkens G, Baets R, Bowers JE. Hybrid integrated platforms for silicon photonics. Materials. 2010;3:1782. https://doi.org/10.3390/ma3031782.
22. Dai D, Liang D, Liu L. Introduction for the integrated photonics: challenges and perspectives feature. Photonics Res. 2015;3:IP1–IP2. https://doi.org/10.1364/PRJ.3.000IP1.
23. Feilchenfeld NB, Nummy K, Barwicz T, Gill D, Kiewra E, Leidy R, Orcutt JS, Rosenberg J, Stricker AD, Whiting C, et al. Silicon photonics and challenges for fabrication. In: Advanced etch technology for nanopatterning VI; SPIE: Bellingham, WA, USA; 2017. p. 101490D.
24. Märki I, Salt M, Herzig HP, Stanley R, El Melhaoui L, Lyan P, Fedeli JM. Characterization of buried photonic crystal waveguides and microcavities fabricated by deep ultraviolet lithography. J Appl Phys. 2005;98: 013103. https://doi.org/10.1063/1.1948507.
25. Gao G, Chen D, Tao S, Zhang Y, Zhu S, Xiao X, Xia J. Silicon nitride O-band (de)multiplexers with low thermal sensitivity. Opt Express. 2017;25:12260–7. https://doi.org/10.1364/OE.25.012260.
26. Shaker LM, Al-Amiery A, Isahak WNRW, et al. Integrated photonics: bridging the gap between optics and electronics for enhancing information processing. J Opt. 2023. https://doi.org/10.1007/s12596-023-01321-8.
27. Zhang J, Xu L, Mao D, D'mello Y, Wei Z, Li W, Plant DV. Temperature-insensitive and low-loss single-mode silicon waveguide crossing covering all optical communication bands enabled by curved anisotropic metamaterial. Nanophotonics. 2023;12:4095–4107. https://doi.org/10.1515/nanoph-2023-0524.

28. Ferreira RAS, André PS, Carlos LD. New photonics trends in sensing (and photovoltaics) towards IOT & e-smart labels. 2022;3(6). https://doi.org/10.1002/adpr.202200133.
29. Fletcher R, Mahindroo A, Patel J. Imperatives for photonics companies in the next wave of growth. Mckinsey & Company. (Imperatives for photonics companies | McKinsey) (2023).
30. Photonics market size, share, forecast | Growth analysis, 2032. https://www.fortunebusinessinsights.com/photonics-market-106525.
31. Rao A, Cyrill M. India's emerging semiconductor ecosystem: key players, India Briefing (2024). https://www.india-briefing.com/news/india-emerging-semiconductor-ecosystem-key-players-34506.html/.
32. Sonawane M, Pawar S, Patil JK, Patil VD. The development of the semiconductor supply chain in India: challenges and opportunities. In: ICT: smart systems and technologies, ICTCS 2023. 2024. pp. 33–42. https://doi.org/10.1007/978-981-99-9489-2_4.
33. India Photonics Market—Forecast (2024—2030), ESR 0510, India photonics market size report, 2021–2026 (industryarc.com).

# Chapter 9
# Innovative Photonics Technologies for Renewable Energy Systems in Industry 5.0

**Peeyush Phogat, Satyam Rawat, Karishma Rawat, Shreya, Ranjana Jha, and Sukhvir Singh**

**Abstract** This chapter examines the potential of novel photonics technologies that could lead to advances in renewable energy systems as an Industry 5.0 case study. This chapter reviews the way state-of-the-art photonic devices and materials are integrated to improve on efficiency, sustainability and functionality of renewable energy solutions. The thematic scope of the journal includes synthesis, processing, and manufacturing methods for developing functional materials from atoms to devices; fabrication processes based on sustainable energy sources that can be used as a representative tool in material design and technology. A further instance deals with photonics solutions for boosting the performance of solar cells, photodetectors, and electrochemical supercapacitors together with a host of recent innovations. It also discusses the challenges and opportunities in incorporating photonic technologies in renewable energy systems, endorsing sustainable growth and environmental wellbeing.

**Keywords** Photonics · Industry 5.0 · Energy harvesting · Photonic materials

## 9.1 Introduction

The emergence of Industry 5.0 marks a transformative leap from the existing framework of Industry 4.0, focusing not only on technological advancement but also on harmonizing human needs with industrial processes. Whereas Industry 4.0 focused on digitization, automation, and machine connectivity via the Internet of Things

---

P. Phogat (✉) · S. Rawat · K. Rawat · Shreya · R. Jha · S. Singh
Research Lab for Energy Systems, Department of Physics, Netaji Subhas University of Technology, New Delhi, India
e-mail: peeyush.phogat@gmail.com

P. Phogat · Shreya
Council of Scientific and Industrial Research–National Institute of Science Communication and Policy Research (CSIR–NIScPR), New Delhi, India

 
V. Bhardwaj et al. (eds.), *Photonics and Optoelectronics in Industry 5.0*, Advances in Optics and Optoelectronics, https://doi.org/10.1007/978-981-96-7594-4_9

(IoT) [1], Industry 5.0 brings the human element back into the center of industrialization [2]. This new phase emphasizes a collaborative relationship between humans and machines, integrating human creativity, decision-making, and empathy into intelligent systems [3]. The ultimate goal of Industry 5.0 is to foster a balance between productivity, sustainability, and societal wellbeing. It also highlights the importance of creating resilient and sustainable industries that can adapt to global challenges, particularly those related to energy consumption [4], climate change [5], and resource management [6]. The shift from Industry 4.0 to Industry 5.0 involves rethinking conventional industrial practices to prioritize sustainability and inclusivity. Human-centered innovation becomes a focal point, where smart industries are not only driven by data but are also aligned with broader environmental and social goals. In this framework, cutting-edge technologies like robotics, artificial intelligence (AI), and the Internet of Things (IoT) continue to be essential, but their application is constrained by sustainability and human empowerment principles [7]. For example, AI-driven systems in Industry 5.0 aim to support, rather than replace, human decision-making, particularly in industries where expertise and creativity are essential. Industry 5.0 simultaneously attempts to solve the urgent problem of environmental sustainability, with the use of clean energy techniques and a decrease in carbon emissions primarily coming from renewable energy sources [8].

Photonics, the science of generating, manipulating, and detecting light, is poised to play an important role in Industry 5.0 by providing cutting-edge solutions that enhance industrial efficiency, precision, and sustainability. As industries evolve toward more sophisticated systems, photonics enables innovations that are central to this transformation. Photonic technologies enable faster, more precise, and energy-efficient industrial activities, ranging from fiber optic communication systems to optical sensors and laser-based manufacturing [9]. The employment of light-based technologies across sectors not only boosts the performance of existing systems but also opens new possibilities for innovation in different areas, including renewable energy, healthcare, telecommunications, and manufacturing. Photonic technologies play a crucial role in Industry 5.0 by serving as a bridge between digital and physical systems, thereby facilitating real-time industrial process monitoring, control, and optimization. Photonic sensors, e.g., provide high-precision measurements in harsh environments [9], while laser-based systems allow for highly accurate material processing. Fiber optic networks offer a robust communication infrastructure that supports the massive data flows required by Industry 5.0's intelligent systems [10]. As industries become more interconnected and reliant on real-time data, photonics offers the speed and efficiency needed to maintain seamless operations while minimizing energy consumption and resource wastage. Furthermore, the versatility of photonic technologies allows them to be integrated into various components of smart manufacturing, from production lines to quality control systems, thereby enhancing overall operational efficiency.

One of the most compelling intersections of photonics and Industry 5.0 lies in the development of renewable energy systems [11]. As the world shifts toward decarbonization, renewable energy has become a critical component of sustainable industrial practices. The integration of photonics into renewable energy systems offers significant potential for improving energy generation, storage, and distribution. Solar energy, in particular, benefits immensely from photonic advancements, where enhanced photovoltaic (PV) systems leverage light manipulation to maximize energy conversion efficiency. Photonics plays a key role in designing solar cells that are more efficient, durable, and cost-effective, thereby contributing to the increased adoption of solar energy in industrial applications [12].

Photonics could revolutionize solar power generation through innovations such as perovskite solar cells, which offer higher efficiency at a lower cost compared to traditional silicon-based cells [13]. Photonic technologies enable better light absorption, enhanced photon management, and reduced energy losses, resulting in solar cells that can operate more effectively under varying environmental conditions. In addition to improving energy conversion in photovoltaic systems, photonics also enhances energy transmission and storage solutions. For instance, fiber optic cables, known for their high data transmission capabilities, can be adapted to transmit energy over long distances with minimal losses, making them an asset in the development of smart energy grids. These grids provide real-time monitoring and optimization of energy distribution, guaranteeing that energy is supplied effectively and sustainably across industrial networks. They are based on optical communication and sensing technologies [14]. Photonics plays a crucial role in renewable energy sources including wind and hydropower, in addition to solar power. To make sure that hydropower facilities and wind turbines run as efficiently as possible, photonic sensors are utilized to track their performance and health [15, 16]. Laser-based systems, e.g., can measure airflow and water flow in real-time, providing critical data that helps improve energy output and reduce operational costs. As renewable energy systems become more integrated with smart grids, photonics will continue to play a pivotal role in ensuring that these systems are not only efficient but also resilient and adaptable to changing environmental conditions. Another promising area where photonics intersects with renewable energy is in the development of energy storage systems [17]. Because renewable energy sources, such as wind and solar power, are sporadic, stable energy supplies require efficient storage methods. By enhancing the performance of batteries and other energy storage technologies, photonics can help with this. For instance, the application of photonic materials in next-generation batteries is being investigated by researchers; this could result in increased energy densities, quicker charging periods, and longer battery lives. Such developments are necessary to facilitate the widespread use of renewable energy in industrial applications, where a steady supply of energy is critical to preserving operational continuity.

Photonic technologies are integral to the evolution of Industry 5.0, offering innovative solutions that enhance the sustainability, efficiency, and precision of modern industrial systems. By integrating photonics into renewable energy systems, industries can achieve their sustainability goals while also improving their operational efficiency. As Industry 5.0 continues to unfold, the role of photonics in renewable energy systems will only become more critical, driving advancements in energy generation, storage, and distribution that align with the broader goals of human-centric and sustainable industrial practices [14]. Through the convergence of photonics and renewable energy, Industry 5.0 promises to create a future where industries are not only smarter but also more attuned to the environmental and social challenges of the modern world.

## 9.2 Industry 5.0: A New Paradigm for Renewable Energy

From the machine-centric, technologically driven world of Industry 4.0, we have witnessed a dramatic evolution with the emergence of Industry 5.0. Industry 5.0 brings back the human component to the industrial equation, whereas Industry 4.0 concentrated mostly on the automation and digitization of industrial processes, highlighting cyber-physical systems, smart factories, and the Internet of Things (IoT) [18, 19]. This change emphasizes human–machine collaboration once again, with a particular emphasis on improving human creativity, decision-making, and well-being. Industry 5.0 places equal emphasis on sustainability, inclusivity, and resilience as it does on technology, in contrast to Industry 4.0. This aligns industrial practices with international initiatives to address environmental concerns such resource depletion, energy inefficiency, and climate change [20]. Industry 5.0 is characterized by its human-centric approach, where intelligent machines and advanced technologies are designed to work alongside humans, complementing rather than replacing human roles. This collaboration aims to create a more harmonious balance between human and machine labor, using technology to enhance human capabilities in creative and complex decision-making processes [15]. One of the most notable shifts is the emphasis on sustainability, where energy efficiency, waste reduction, and environmental responsibility are integrated into industrial practices. Industry 5.0 envisions a future where industries not only strive for economic productivity but also actively contribute to the health of the planet and the wellbeing of society. In this sense, the key distinction between Industry 4.0 and Industry 5.0 lies in the latter's holistic approach, which values human welfare, environmental sustainability, and ethical considerations alongside technological advancement [16].

Central to this new industrial paradigm is the transition to sustainable energy solutions. Industry 5.0 understands how critical it is to shift away from fossil fuels and toward renewable energy sources like solar, wind, hydro, and bioenergy. Achieving climate goals, cutting greenhouse gas emissions, and guaranteeing future generations have access to clean energy depend on this change [21]. The integration of renewable energy systems into industrial processes is no longer just a matter of corporate responsibility but a key driver of operational efficiency and long-term competitiveness. With renewable energy technologies becoming more advanced and cost-effective, industries are now able to integrate clean energy solutions into their operations, minimizing their carbon footprint while reducing energy costs. One of the central pillars of Industry 5.0's sustainability agenda is the adoption of circular economy principles. Instead of the traditional linear model of resource extraction, production, and disposal, Industry 5.0 encourages industries to adopt circular practices that involve reusing, recycling, and regenerating resources [22]. This shift is especially crucial in the energy sector, where minimizing waste and maximizing the efficiency of energy systems can have a profound impact on overall sustainability. The integration of renewable energy sources, combined with energy storage solutions and smart grid systems, allows industries to operate in a more energy-efficient manner, reducing waste and minimizing their environmental impact.

Advanced technologies like artificial intelligence (AI), the Internet of Things (IoT), and photonics, which are combining to develop intelligent energy management systems, are at the core of Industry 5.0's energy transformation. By evaluating enormous volumes of data and generating judgments in real time that improve energy efficiency, artificial intelligence (AI) plays a critical role in optimizing energy use. In an industrial setting, AI can predict energy demand, identify inefficiencies in energy usage, and recommend adjustments that reduce waste and lower costs [23]. At the heart of Industry 5.0's energy transformation is advanced technologies such as artificial intelligence (AI), the Internet of Things (IoT), and photonics, which are converging to create intelligent energy management systems. AI plays a crucial role in optimizing energy consumption by analyzing vast amounts of data and making real-time decisions that enhance energy efficiency.

Conversely, the Internet of Things (IoT) offers the infrastructure required for smooth communication between gadgets, sensors, and systems, making it possible to monitor and manage energy consumption in real time [18]. In Industry 5.0, IoT sensors are deployed across industrial facilities to track energy consumption patterns, monitor equipment performance, and detect inefficiencies or malfunctions. This data is then fed into AI systems, which use it to make informed decisions about energy management. By connecting renewable energy systems with smart grids and industrial operations, IoT allows for more dynamic and responsive energy management, ensuring that energy is used more efficiently and sustainably [19]. Photonics, the science of using light-based technologies, is another key enabler of intelligent energy management in Industry 5.0. Photonic technologies play a vital role in improving the performance of renewable energy systems, from enhancing the efficiency of solar panels to enabling high-speed optical communication networks for energy management systems. Photonics is particularly important in solar energy, where advanced photovoltaic (PV) systems use photonic materials and techniques to capture and convert sunlight into electricity more efficiently. Additionally, photonic sensors provide high-precision monitoring of energy systems, allowing industries to detect energy losses, optimize performance, and ensure the long-term reliability of renewable energy installations. Furthermore, fiber optic communication systems, a key component of photonics, enable the rapid transmission of data across smart grids, ensuring that energy generation, storage, and consumption are all synchronized in real-time [20]. The convergence of AI, IoT, and photonics is revolutionizing how energy is managed in industrial settings, making it possible to integrate renewable energy sources more effectively into the energy mix. Smart grids, powered by AI and IoT, allow for real-time balancing of energy supply and demand, ensuring that renewable energy is utilized efficiently and that energy storage systems are deployed when needed [24]. Photonic sensors and communication networks further enhance the efficiency of these systems, enabling industries to monitor energy usage with unprecedented precision and adjust operations accordingly.

## 9.3 Photonic Technologies in Renewable Energy Systems

Photonics, the science, and technology of light, is driving significant innovation in energy systems by enhancing the efficiency and performance of energy generation, storage, and distribution. As industries and societies strive for cleaner and more sustainable energy sources, photonic technologies offer promising solutions across various sectors. In energy generation, particularly in solar power, photonics plays a pivotal role. Advanced photovoltaic (PV) cells leverage photonic materials and techniques to capture and convert sunlight into electricity more efficiently [25]. Innovations such as multi-junction solar cells and perovskite-based systems are optimizing light absorption, significantly improving energy conversion rates. Through the improvement of energy storage system materials and the enhancement of charging and discharging procedures through accurate light-based monitoring, photonics helps to advance battery technologies in the energy storage industry, notably next-generation lithium-ion, and solid-state batteries [26]. These developments lengthen battery life, boost energy density, and speed up charging.

Photonics also plays a crucial role in energy distribution through fiber optic networks, which enable high-speed, low-loss transmission of data necessary for managing energy flows in smart grids. This lowers energy losses and ensures the effective use of renewable energy sources by enabling real-time monitoring and optimization of energy distribution across power networks. Furthermore, photonic sensors provide precise measurements in harsh environments, improving the reliability of energy systems. By enhancing the generation, storage, and distribution of energy, photonics is at the forefront of innovation in sustainable energy systems [21], making it a key enabler of the renewable energy transition in Industry 5.0.

### *9.3.1 Key Photonic Technologies*

(a) **Solar Photovoltaics**: Solar photovoltaics (PV) is perhaps the most prominent example of how photonics is revolutionizing energy generation. Traditional solar cells have limitations in their ability to capture and convert sunlight into electricity efficiently. However, recent advances in light absorption and photon management have significantly improved the efficiency and performance of PV systems [27]. The creation of multi-junction solar cells, which are intended to absorb various sun wavelengths, is one significant advance. These cells use layers of semiconducting materials that each absorb a specific range of the solar spectrum, maximizing energy conversion [28]. Moreover, photon management techniques, such as light-trapping strategies, have been introduced to minimize the loss of light that typically escapes from the surface of solar cells. By utilizing nanostructures or textured surfaces, light can be captured more effectively, thereby increasing the overall conversion efficiency [29]. Materials like perovskite, which have excellent light absorption properties, are also showing great promise in next-generation solar cells, offering higher energy conversion rates at lower production costs. Photonic crystals and plasmonic are being employed to manipulate light at the nanoscale, ensuring that photons are absorbed and utilized to their fullest potential, thus driving innovation in solar power generation.

(b) **Laser-Based Energy Transfer**: Laser-based energy transfer is an emerging photonic technology that holds significant potential for applications where traditional wired energy transmission is not feasible. This non-contact energy transfer method uses highly focused beams of light (usually lasers) to transmit energy across long distances without the need for physical connections [30]. Such technology is particularly useful in environments where cables are impractical, such as in remote locations or space applications. Lasers can deliver energy with high precision to targeted devices or systems, making this method ideal for powering drones [31], satellites [32], or remote sensors [33]. In addition to the potential for long-distance energy transfer, laser-based systems can operate in environments where traditional energy transmission methods are susceptible to damage or interference. This capability could open new avenues for energy delivery in harsh or inaccessible environments, further enhancing the flexibility and efficiency of renewable energy systems as shown in Fig. 9.1.

(c) **Optical Sensors**: Photonics also plays a vital role in energy systems using optical sensors. These sensors rely on light to detect and measure physical properties such as temperature, pressure, and strain with high precision and sensitivity [34]. In renewable energy systems, optical sensors are deployed to monitor the performance of critical components, such as wind turbines, solar panels, and batteries. Their ability to operate in harsh environments, such as extreme temperatures or electromagnetic interference, makes them invaluable for ensuring the reliability and efficiency of energy systems [35]. Optical fiber sensors offer real time, remote monitoring capabilities, which is crucial for predictive maintenance in renewable energy installations. By continuously tracking the condition of energy systems, optical sensors help prevent costly downtimes and inefficiencies. For example, in solar farms, these sensors can measure sunlight intensity and the condition of photovoltaic panels, allowing operators to optimize power generation and detect potential issues before they escalate [36].

(d) **Fiber Optics for Energy Transmission**: Fiber optics, traditionally used in telecommunications, are now finding applications in energy transmission systems. Unlike electrical transmission lines, which lose energy over long distances due to resistance and heat, fiber optics can transmit light-based energy with minimal loss. This technology is being explored for transmitting energy in the form of laser beams or other light-based carriers over significant distances [37]. By utilizing optical fibers, energy can be delivered more efficiently and with greater speed, reducing transmission losses and enhancing the overall efficiency of energy grids. Fiber optic systems are also integral to smart grids, where they enable rapid communication between energy generation, storage, and consumption points. The high-speed data transmission capabilities of fiber optics allow energy systems to be monitored and managed in real time, ensuring

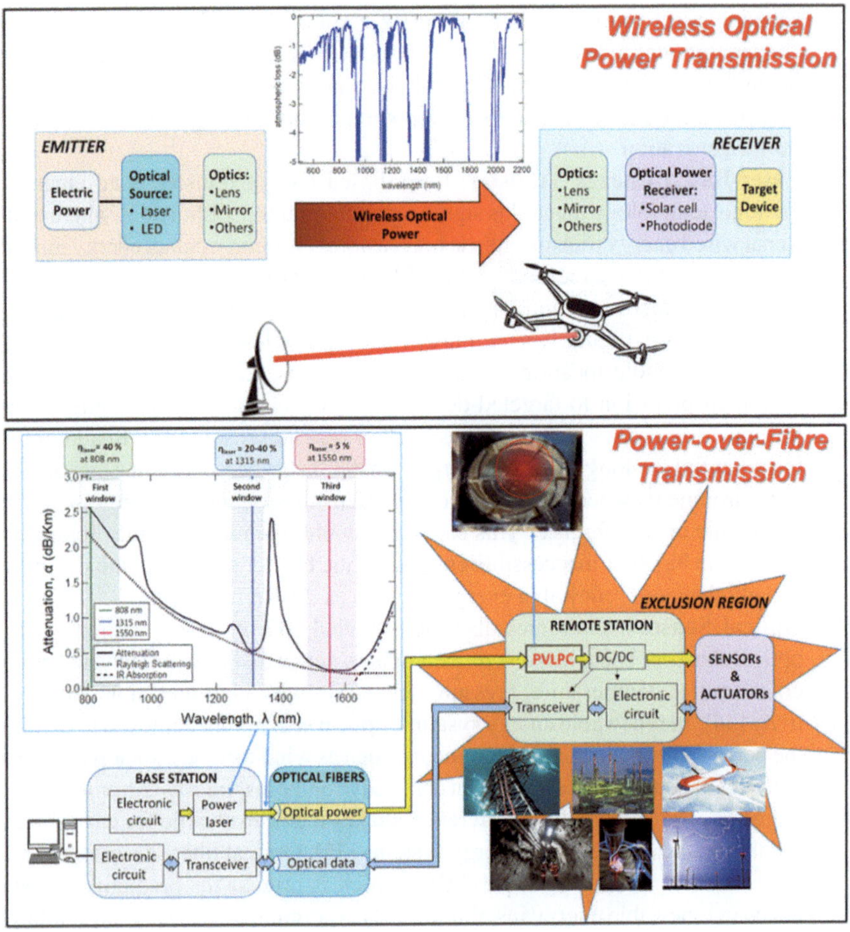


**Fig. 9.1** The image depicts two methods of optical power transmission: wireless optical power transmission, where a focused optical beam from a laser or LED transfers energy wirelessly to a receiver, and power-over-fiber transmission, which uses optical fibers to transmit both power and data with minimal signal loss. These technologies are vital for powering remote systems, especially in hazardous or hard-to-reach environments. This image is reproduced with permission from Ref. [26] © 2021 Elsevier Inc

that renewable energy resources are utilized optimally [38]. This precise control and communication are essential for balancing supply and demand, particularly in grids that integrate intermittent energy sources like solar and wind power.

## 9.4 Innovative Photonics Applications in Renewable Energy

### 9.4.1 Photonics in Solar Energy

Perovskite solar cells are a top contender for next-generation solar technologies because of their remarkable photon absorption and conversion efficiency, which has attracted a lot of interest in recent years. Perovskite materials, a class of chemicals with exceptional light-harvesting capabilities over a wide range of the solar spectrum, possess distinctive features that are crucial to their success [39]. Perovskite solar cells, in contrast to conventional silicon-based solar cells, can be adjusted to more efficiently absorb light wavelengths, enabling them to collect more sunlight and convert it into electricity [40].

One of the major advantages of perovskites is their ability to be processed using low-cost, solution-based techniques, which can be applied to flexible substrates. This makes them not only highly efficient but also inexpensive to manufacture compared to traditional photovoltaic technologies. Furthermore, advances in photon management techniques, such as integrating photonic crystals and plasmonic nanoparticles into the perovskite structure, have further enhanced their light absorption capabilities [41]. These innovations allow perovskite solar cells to trap and utilize photons more effectively, reducing energy losses and improving overall power conversion efficiency. With ongoing research focusing on improving the stability and longevity of perovskite materials, this technology holds tremendous promise for transforming solar energy into a more viable, efficient, and cost-effective solution for widespread deployment. Another potent technique that makes use of photonics to improve energy generation is concentrated solar power (CSP), as illustrated in Fig. 9.2. CSP works by focusing sunlight onto a tiny area using lenses or mirrors, creating high heat that is subsequently converted into electrical power. Unlike conventional photovoltaic systems that directly convert sunlight into electricity, CSP relies on the thermal energy of concentrated sunlight [42]. This concentrated energy is typically used to heat a fluid, such as molten salt or water, which then drives a turbine to generate power.

Photonics plays a vital role in CSP systems through the precise design and arrangement of mirrors and lenses, which maximize the concentration of sunlight onto the receiver. By using parabolic mirrors, heliostats, or Fresnel lenses, CSP systems can concentrate sunlight by a factor of hundreds, significantly increasing the thermal energy available for electricity production [44]. Because they can incorporate thermal energy storage systems and function even in the absence of sunlight, CSP systems are especially beneficial in areas with significant solar insolation. As a result, CSP offers a reliable and scalable renewable energy solution, capable of supporting large-scale energy demands.

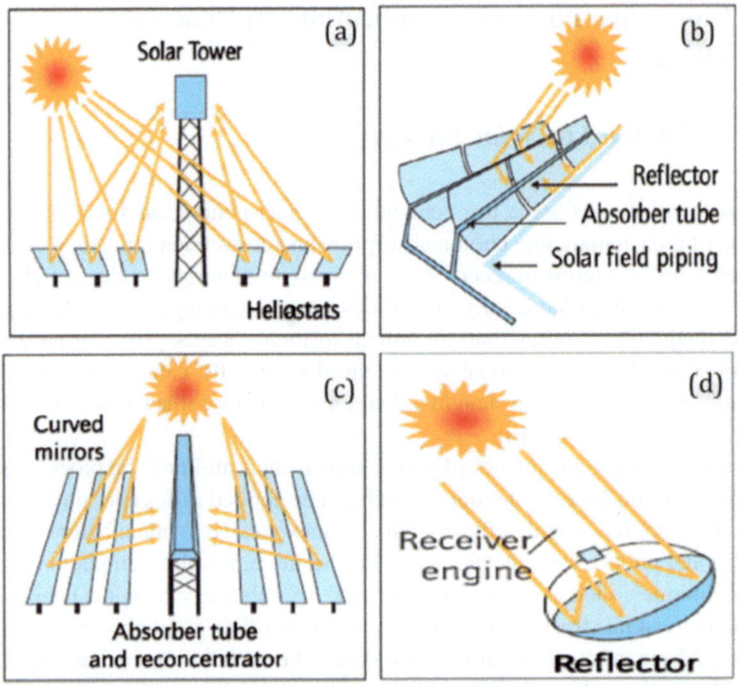


**Fig. 9.2** Various solar thermal energy storage technologies. **a** Solar tower with heliostats focuses sunlight onto a central receiver for thermal energy storage. **b** Parabolic trough system uses reflectors to concentrate sunlight onto an absorber tube for heat capture and storage. **c** Linear Fresnel reflectors focus sunlight onto an absorber tube with curved mirrors for energy collection. **d** Parabolic dish system concentrates sunlight onto a receiver/engine, where thermal energy is stored and converted into mechanical or electrical energy. This image is reproduced with permission from Ref. [43] Copyright © 2014 Elsevier B.V

## 9.4.2 Photonics in Wind and Hydroelectric Energy

Optical sensors have become indispensable for optimizing the performance of renewable energy systems, particularly in solar power plants, wind farms, and battery storage facilities. These sensors use light to very precisely measure physical factors like strain, temperature, pressure, and vibration. In solar power systems, optical sensors monitor sunlight intensity and the performance of photovoltaic (PV) panels in real time [45], allowing operators to optimize energy output by adjusting the orientation of solar panels or diagnosing faults before they affect overall efficiency. In wind turbines, optical fiber sensors are embedded within turbine blades and towers to continuously measure mechanical stress, strain, and vibration, providing critical data for predictive maintenance [46]. This helps prevent equipment failure and minimizes downtime, ensuring that turbines operate at peak efficiency [47]. Optical sensors are also highly resistant to environmental challenges such as extreme temperatures, electromagnetic interference, and harsh weather conditions. This resilience makes them particularly well-suited for use in remote or harsh environments, where renewable energy installations are often located. Moreover, their real-time monitoring capabilities allow for instantaneous adjustments to energy systems, maximizing efficiency and minimizing waste. By leveraging optical sensors, renewable energy systems can achieve greater operational longevity, lower maintenance costs, and enhanced reliability, all of which are crucial for the continued adoption of clean energy solutions [48].

Laser-based flow diagnostics is a cutting-edge technique used to analyze the behavior of fluids, particularly in wind turbines and water flow systems. In wind turbines, lasers are used to measure airflow patterns around turbine blades, enabling precise adjustments to the blade angle and orientation for maximum energy capture. This technique, known as Laser Doppler Anemometry (LDA), allows operators to visualize wind flow in real-time and identify areas where turbulence or inefficiencies may occur [49]. By optimizing the aerodynamic performance of wind turbines through laser diagnostics, energy output can be significantly improved, while wear and tear on turbine blades are minimized. Similarly, in water flow systems, lasers are employed to measure the velocity and turbulence of water in hydropower plants, dams, or other water-based energy systems as shown in Fig. 9.3. This non-invasive, high-precision diagnostic tool provides valuable insights into how water flows through turbines, pipes, and channels, enabling operators to fine-tune the system for optimal performance. The use of laser diagnostics in both wind and water systems allows for greater efficiency, reduced operational costs, and enhanced sustainability, making it a crucial technology for the future of renewable energy.

### *9.4.3 Photonics in Smart Grids*

Fiber optic technology has become indispensable for high-speed communication within modern energy grids, enabling rapid data transmission with minimal signal loss. Fiber optics, which use light instead of conventional copper cables to carry data over great distances at extremely fast speeds [50], are perfect for the complex requirements of energy management in Industry 5.0. These systems are particularly vital in smart grids, where real-time data on energy generation, storage, and consumption must be transmitted and processed instantly to ensure optimal performance. Energy systems can quickly exchange information via fiber optics, modifying power allocation in response to variations in renewable energy sources' supply and demand, such as wind and solar power. Advanced technologies that rely on the quick and seamless transmission of massive amounts of data, including artificial intelligence (AI) and the Internet of Things (IoT), can also be integrated thanks to fiber optics' high bandwidth and low latency [51]. In short, fiber optics enable smart grids to operate more efficiently, ensuring reliable and balanced energy distribution [40].

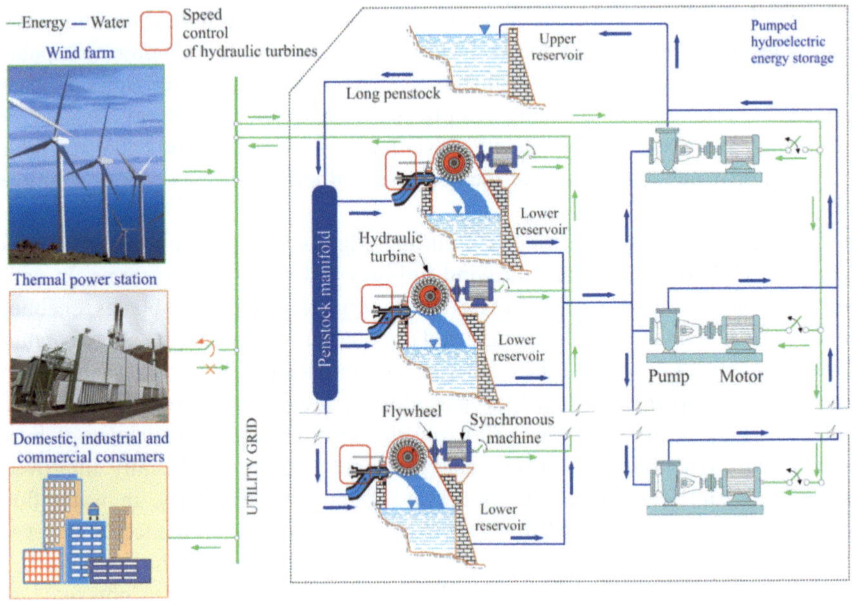


**Fig. 9.3** Schematic of a pumped hydroelectric energy storage (PHES) system integrated with renewable and conventional energy sources. Energy from wind farms, thermal power stations, and the utility grid is used to pump water from lower to upper reservoirs via motor-driven pumps during periods of excess energy generation. During high demand, water flows back down through hydraulic turbines, generating electricity. Flywheels and synchronous machines provide speed control and energy balancing, enhancing the system's efficiency and stability for domestic, industrial, and commercial consumers. This image is reproduced with permission from Ref. [39] © 2023 by the authors. Licensee MDPI, Basel, Switzerland

Conversely, photonic sensors are essential for tracking the grid's efficiency and health in real time. These sensors provide extremely sensitive and accurate measurements by using light to detect changes in physical characteristics such as temperature, pressure, and strain across energy systems. In smart grids, photonic sensors are deployed at key points to monitor the condition of equipment, power lines, and renewable energy installations, offering insights into performance and detecting potential failures [52]. For example, in wind farms or solar arrays, photonic sensors can track operational efficiency and environmental conditions, sending real-time data to central control systems. This allows for predictive maintenance, where potential issues can be addressed before they result in costly downtimes or energy losses. Moreover, photonic sensors are immune to electromagnetic interference [53], making them especially suited for harsh or high-voltage environments. Through their precision and resilience, these sensors enhance the reliability and safety of modern energy grids, ensuring continuous operation and optimal efficiency. Together, fiber optics and photonic sensors are essential for creating smarter, more responsive, and efficient energy infrastructures.

## 9.5 Integration of Photonics with Emerging Technologies in Energy Systems

By combining with cutting-edge technologies like artificial intelligence (AI), the Internet of Things (IoT), and energy storage systems, photonics is revolutionizing contemporary energy systems [54]. This convergence is driving innovation, enhancing efficiency, and enabling more sustainable energy solutions. Below, we explore how photonics integrates with these emerging technologies to optimize energy production, distribution, and storage.

### *9.5.1 Photonics and Artificial Intelligence (AI)*

The fusion of photonics with artificial intelligence (AI) is creating new avenues for energy generation and consumption optimization. AI-powered photonic devices can process large volumes of data at high speeds using light-based systems, which are significantly faster and more efficient than their electronic counterparts [55]. In the context of renewable energy, AI algorithms are employed to analyze data from photonic sensors embedded in solar panels, wind turbines, and energy grids. By learning patterns and predicting energy generation from renewable sources, AI can dynamically adjust energy production based on real-time demand and weather conditions. For example, in solar power plants, AI-powered photonic systems can predict periods of high and low sunlight, allowing for the optimal adjustment of energy capture and storage [56]. Similarly, wind turbines equipped with AI-driven photonic sensors can adjust their orientation to maximize energy production, even in fluctuating wind conditions. AI's ability to predict energy demand patterns also enhances grid management. Using photonic technology to monitor the grid in real time, AI systems can automatically divert energy to areas with higher demand, reducing wastage and ensuring a balanced, efficient energy flow. This degree of accuracy and optimization opens the door for Industry 5.0's smarter, more adaptive grids by lowering operating costs and improving the sustainability and dependability of renewable energy systems [57].

### *9.5.2 Photonics and Internet of Things (IoT)*

Incorporating photonics with the Internet of Things (IoT) is essential for managing and monitoring energy systems in real time [58]. IoT-enabled photonic sensors are deployed throughout energy networks, collecting vital data on system performance, environmental conditions, and equipment health. These sensors communicate this data wirelessly to central management systems, where it is analyzed to optimize energy use and enhance grid stability [59]. For example, in a smart grid, photonic sensors can continuously monitor the condition of power lines, detecting any fluctuations in voltage, temperature, or pressure. This data is relayed via IoT networks to control centers, where operators or automated systems can make quick adjustments to prevent equipment failure or energy loss. Photonic sensors are highly sensitive, capable of detecting minor changes in the system, making them ideal for predictive maintenance. This prevents breakdowns and minimizes downtime by addressing issues before they escalate. In renewable energy applications, IoT-enabled photonic devices also monitor environmental factors that impact energy generation [60]. Solar farms, for instance, use photonic sensors to track sunlight intensity and panel efficiency, while wind farms monitor wind speed and turbine performance in real-time. With IoT, this data is sent to the cloud, allowing operators to make data-driven decisions, even from remote locations. To integrate intermittent renewable energy sources into the grid and provide a consistent and dependable energy supply, this degree of responsiveness and connectedness is essential. Photonics, in conjunction with IoT, allows for more intelligent, connected energy systems capable of self-monitoring and self-adjustment, improving overall energy efficiency [61].

### *9.5.3 Photonics and Energy Storage Systems*

Energy storage is a critical component of renewable energy systems, and photonic technologies are significantly improving the performance of energy storage systems, particularly batteries as shown in Fig. 9.4. In energy storage systems, photonics is used to enhance the materials and processes involved in storing and releasing energy, leading to improvements in battery efficiency, energy density, and lifespan [62]. Solid-state batteries, which are seen as the next generation of energy storage technologies, are one area where photonics is having a significant influence. Solid-state batteries offer higher safety and energy density by substituting a solid electrolyte for the liquid or gel electrolyte used in conventional batteries. Photonic techniques, such as laser-assisted manufacturing and nanophotonic material structuring, are used to optimize the solid-state battery's components, enabling faster ion transport and reducing energy losses [63]. These developments lead to batteries that are more dependable and efficient for use in grid energy storage and electric car applications because they can store more energy, charge more quickly, and last longer.

Moreover, photonic sensors integrated into energy storage systems enable precise monitoring of battery health in real time [50]. These sensors can detect changes in temperature, pressure, and chemical composition, allowing for early detection of potential issues such as overheating or degradation. By continuously monitoring the battery's condition, photonic technologies contribute to extending the battery's lifespan and preventing failures, which is essential for the long-term reliability of renewable energy storage solutions. Another innovative application of photonics in energy storage is in light-assisted charging [51]. Certain advanced battery materials, such as lithium-sulfur or graphene-based systems, exhibit improved charging efficiency when exposed to specific wavelengths of light. Photonic devices can be used

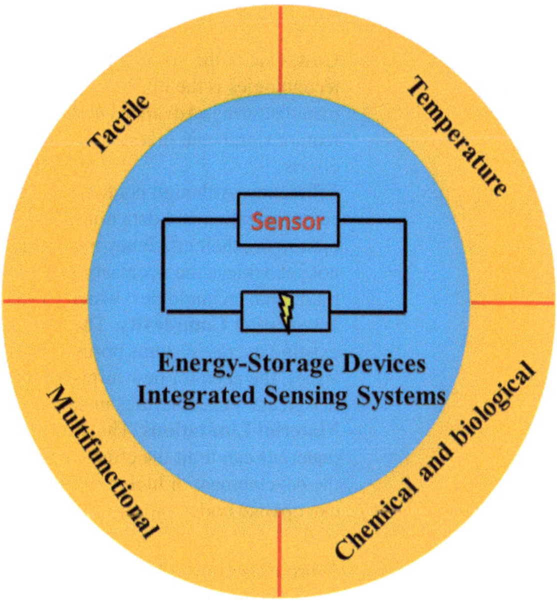


**Fig. 9.4** Energy storage devices integrated with sensors for tactile, temperature, and chemical/biological detection, enabling real-time monitoring in wearable electronics. This image is reproduced with permission from Ref. [64]. 

to enhance the charging process, reducing the time it takes to charge batteries while maintaining high energy density. This opens possibilities for faster and more efficient charging of EVs and portable electronic devices, contributing to the widespread adoption of renewable energy technologies [65].

## 9.6 Challenges and Opportunities

Here's a structured overview of the **Challenges and Opportunities** related to the integration of photonics in energy systems in Industry 5.0, presented in table form.

| Aspect | Description |
|---|---|
| Technical challenges | – **Cost**: One of the primary barriers to the widespread adoption of photonic technologies is the high cost associated with their development and manufacturing. Advanced materials and fabrication techniques often require significant investment, which can hinder commercialization efforts<br>– **Efficiency**: Although photonic technologies offer high-speed communication and data transmission, there are still challenges in optimizing their efficiency for energy applications. Current systems may not yet achieve the necessary performance levels to compete with traditional technologies, which can slow down adoption<br>– **Integration Complexity**: The integration of photonic technologies with existing energy systems poses significant challenges. Compatibility issues, the need for new infrastructure, and the complexity of system design can create hurdles in adopting photonics for energy management<br>– **Material Limitations**: The availability and performance of photonic materials can limit the effectiveness of certain applications. For instance, the development of high-performance photonic crystals or sensors that can operate under various environmental conditions remains a challenge [66] |
| Scalability | – **From Research to Deployment**: Bridging the gap between laboratory research and industrial-scale deployment is a critical challenge. While many photonic technologies show promise in controlled environments, demonstrating their effectiveness in real-world applications remains a significant hurdle<br>– **Standardization**: The lack of standardized protocols and metrics for assessing photonic technologies can complicate efforts to scale these solutions. The energy sector often relies on established standards, and the introduction of new technologies requires consensus on performance metrics, safety, and interoperability<br>– **Investment and Funding**: Transitioning from research to commercial viability necessitates substantial investment. Attracting funding and investment for pilot projects and large-scale deployments is essential but can be challenging, especially for new technologies that require time to prove their return on investment |

(continued)

(continued)

| Aspect | Description |
|---|---|
| Opportunities for innovation | – **Enhanced Energy Generation**: Photonic technologies can significantly boost the efficiency of solar energy systems through advanced materials and designs, leading to higher energy yields. Innovations such as multi-junction solar cells and light-trapping structures offer exciting prospects for energy generation[54]<br>– **Smart Grid Solutions**: Smart grids might be completely transformed by the incorporation of photonic sensors and communication systems, which would allow for effective energy management, predictive maintenance, and real-time monitoring. Energy systems that are more flexible and able to adjust to shifting supply and demand conditions may result from this<br>– **Energy Storage Advancements**: Photonics can enhance the performance of energy storage systems, such as batteries, by enabling faster charging, improved efficiency, and extended lifespans through innovative materials and manufacturing processes<br>– **Innovative Applications in Transportation**: The integration of photonic technologies in electric vehicles can optimize energy management, enhance charging infrastructure, and improve vehicle-to-grid communication, fostering a more sustainable transportation ecosystem<br>– **Cross-Sector Collaborations**: There is significant potential for cross-industry collaboration, particularly between photonics and sectors like telecommunications, automotive, and manufacturing, leading to innovative solutions that leverage strengths across industries for sustainable energy systems |

## 9.7 Future Prospects and Conclusion

The landscape of photonics in energy systems is rapidly evolving, driven by groundbreaking trends that promise to revolutionize how we generate, store, and manage energy. One of the most exciting areas is photonic quantum technologies [67]. These technologies leverage the principles of quantum mechanics to develop highly efficient energy systems. Quantum photonics can enhance the performance of solar cells and energy storage systems by enabling unprecedented levels of light manipulation and energy conversion efficiency. For instance, quantum dots are being researched for their ability to improve light absorption in solar panels, potentially leading to higher energy yields [68]. Bio-photonics in energy harvesting is another emerging trend that merges biological processes with photonic technologies [69]. This field explores the use of light-based techniques for energy harvesting from biological sources. For example, researchers are investigating how to capture energy from photosynthetic organisms, such as algae, using advanced photonic materials and structures. This innovative approach can lead to sustainable energy solutions that harness natural processes, minimizing the environmental impact associated with traditional energy generation methods. Additionally, the development of new materials for photonic energy systems is crucial for advancing efficiency and performance. Materials such as perovskites and 2D materials (e.g., graphene) are at the forefront of this research. These materials exhibit exceptional light absorption and emission properties, paving the way for more efficient solar cells, photodetectors, and light-emitting devices [58, 70]. The integration of these novel materials into energy systems can significantly enhance their performance and reliability.

To fully utilize the promise of these technologies, several research gaps in photonics for energy applications still need to be filled, notwithstanding recent developments in the field. One critical area is the integration of photonic systems with existing energy infrastructures [71]. Research is needed to explore how photonic technologies can seamlessly interact with traditional energy systems, ensuring compatibility and enhancing overall efficiency. Moreover, the long-term stability and durability of photonic devices in energy applications require further investigation. Maintaining the longevity and dependability of these systems requires an understanding of how they operate over lengthy stretches of time and in a variety of environmental circumstances. Additionally, the economic feasibility of large-scale implementation of photonic technologies needs to be assessed. This includes conducting cost–benefit evaluations and investigating business models that can enable the implementation of these technologies in the energy sector.

Global adoption of sustainable energy systems could be greatly accelerated by cutting-edge photonic technologies. Photonics can play a crucial role in accomplishing global renewable energy targets by improving energy storage options, increasing the efficiency of solar energy output, and enabling real-time monitoring and management of energy systems. For instance, as governments attempt to achieve their commitments under international agreements like the Paris Agreement, the integration of sophisticated photonic technologies might assist reduce greenhouse gas emissions and dependency on fossil fuels. Furthermore, the scalability of photonic solutions means that they can be deployed in diverse applications, from residential solar energy systems to large-scale solar farms. This versatility allows for broader access to renewable energy, contributing to energy equity and security across regions. By facilitating a transition to more sustainable energy practices, photonics can help create a resilient and adaptive energy infrastructure that meets the needs of future generations.

In summary, the integration of photonics into renewable energy systems represents a transformative shift toward sustainable energy solutions in the context of Industry 5.0 [72]. The emerging trends in photonic technologies—such as photonic quantum technologies, bio-photonics, and new materials—underscore the importance of this field in driving innovation and improving the efficiency of energy systems. However, addressing the existing research gaps is crucial for unlocking the full potential of photonics in energy applications. The impact of these innovations on global renewable energy goals cannot be overstated. By enhancing energy generation, storage, and management, photonic technologies offer a pathway to accelerate the transition toward sustainable energy systems worldwide. The future outlook for photonic innovations in energy is promising, with the potential to play a central role in building resilient, efficient, and sustainable energy infrastructures that benefit society and the planet [21]. As we move forward, collaboration between researchers, industry stakeholders, and policymakers will be vital in fostering the development and adoption of photonic technologies. By embracing these innovations, we can create a future where renewable energy is not only abundant but also accessible and sustainable, aligning with the broader goals of Industry 5.0.

## References

1. Alaba FA, Sani U, Dada EG, Mohammed BH. A IoT-enabled smart grids: advancing energy efficiency and renewable energy integration. In: Lecture notes on data engineering and communications technologies. 2024; 192. pp. 59–79. https://doi.org/10.1007/978-3-031-53433-1_4; (PDF) Industry 5.0: Deepening the Subject. (n.d.).
2. Algora C, García I, Delgado M, Peña R, Vázquez C, Hinojosa M, Rey-Stolle I. Beaming power: photovoltaic laser power converters for power-by-light. Joule. 2022;6(2):340–68.
3. Asilian A, Shahinzadeh H, Zanjani SM, Gharehpetian GB, Abdelaziz AY, Tanwar S. The role of microelectronics for smart cities, smart grids and industry 5.0: challenges, solutions, and opportunities. In: 2023 13th smart grid conference, SGC 2023. https://doi.org/10.1109/SGC61621.2023.10459310.
4. Beheshti Asl M, Fofana I, Meghnefi F. Review of various sensor technologies in monitoring the condition of power transformers. Energies. 2024;17(14):3533. https://doi.org/10.3390/EN17143533.
5. Ogundipe OB, Okwandu AC, Abdulwaheed SA. Recent advances in solar photovoltaic technologies: efficiency, materials, and applications. GSC Adv Res Rev. 2024;20(1):159–75. https://doi.org/10.30574/gscarr.2024.20.1.0259.
6. Butt MA, Mateos X, Piramidowicz R. Photonics sensors: a perspective on current advancements, emerging challenges, and potential solutions. Phys Lett A. 2024;516: 129633.
7. Chaudhry AU, Lamontagne G, Yanikomeroglu H. Laser intersatellite link range in free-space optical satellite networks: impact on latency. IEEE Aerosp Electron Syst Mag. 2023;38(4):4–13. https://doi.org/10.1109/MAES.2023.3241142.
8. Chen L, Fang B, Ke Q, Wangyang P, Hu K, Zhang W. Quasi-hemispherical pit array textured surface for increasing the efficiency of thin-film solar cells. AIP Adv. 2022;12(1). https://doi.org/10.1063/5.0074759/2819055.
9. Chen Y, Ni YQ, Ye XW, Yang HX, Zhu S. Structural health monitoring of wind turbine blade using fiber Bragg grating sensors and fiber optic rotary joint. 2012;8345:885–890. https://doi.org/10.1117/12.914961.
10. Coetzer J, Kuriakose RB, Vermaak HJ. Using the OEE score to enable collaborative decision-making for human–machine interaction in an Industry 5.0 setting. In: World conference on information systems for business management. Singapore: Springer Nature Singapore; 2023. pp. 279–292.
11. Davies R, Coole T, Smith A. Review of socio-technical considerations to ensure successful implementation of Industry 4.0. Procedia Manuf. 2017;11:1288–95. https://doi.org/10.1016/J.PROMFG.2017.07.256.
12. Dong X, Chen X, Jiang X, Yang N. Light-assisted energy storage devices: principles, performance, and perspectives. Adv Energy Mater. 2023;13(38):2301143. https://doi.org/10.1002/AENM.202301143.
13. Fakharuddin A, Shabbir U, Qiu W, Iqbal T, Sultan M, Heremans P, Schmidt-Mende L. Inorganic and layered perovskites for optoelectronic devices. Adv Mater. 2019;31(47):1807095. https://doi.org/10.1002/ADMA.201807095.
14. Fermann ME, Galvanauskas A, Sucha G, Harter D. Fiber-lasers for ultrafast optics. Appl Phys B. 1997;65:259–75.

15. Feuermann D, Gordon JM. Solar fiber-optic mini-dishes: a new approach to the efficient collection of sunlight. Sol Energy. 1999;65(3):159–70.
16. Fontaine BL, Vidal F, Jiang Z, Johnston TW, Fontaine BL, Jiang Z, Chien CY, Comtois D, Desparois A, Johnston TW, Kieffer J-C, Pépin H, Mercure HP. Filamentation of ultrashort pulse laser beams resulting from their propagation over long distances in air. Pubs Aip Org. 1999;6: 123116. https://doi.org/10.1063/1.873715.
17. Fareq M, Fitra M, Irwanto M, Syafruddin HS, Gomesh N, Farrah S, Rozailan M. Solar wireless power transfer using inductive coupling for mobile phone charger. In: Proceedings of the 2014 IEEE 8th international power engineering and optimization conference, PEOCO 2014. pp. 473–476. https://doi.org/10.1109/PEOCO.2014.6814475.
18. Ghobakhloo M, Fathi M, Iranmanesh M, Vilkas M, Grybauskas A, Amran A. Generative artificial intelligence in manufacturing: opportunities for actualizing Industry 5.0 sustainability goals. J Manuf Technol Manage. 2024;35(9):94–121. https://doi.org/10.1108/JMTM-12-2023-0530/FULL/HTML.
19. Gong Z, Xiang Z, OuYang X, Zhang J, Lau N, Zhou J, Chan CC. Wearable fiber optic technology based on smart textile: a review. Materials 2019;12(20):3311. https://doi.org/10.3390/MA1220 3311.
20. Green MA, Keevers MJ, Thomas I, Lasich JB, Emery K, King RR. 40% efficient sunlight to electricity conversion. Prog Photovoltaics Res Appl. 2015;23(6):685–91. https://doi.org/10. 1002/PIP.2612.
21. Satheesh S, Shukla K. Industry 5.0, quality of life, and sustainable adaptation to climate change: challenges and the transformational way forward. In: quality of life and climate change: impacts, sustainable adaptation, and social-ecological resilience. IGI Global; 2024. pp. 200–219.
22. Gangwar RK, Kumari S, Pathak AK, Gutlapalli SD, Meena MC. Optical fiber based temperature sensors: a review. Optics. 2023;4(1):171–97.
23. Ghaffour N, Bundschuh J, Mahmoudi H, Goosen MFA. Renewable energy-driven desalination technologies: a comprehensive review on challenges and potential applications of integrated systems. Desalination. 2015;356:94–114. https://doi.org/10.1016/J.DESAL.2014.10.024.
24. Hossain MS, Madlool NA, Rahim NA, Selvaraj J, Pandey AK, Khan AF. Role of smart grid in renewable energy: an overview. Renew Sustain Energy Rev. 2016;60:1168–84. https://doi. org/10.1016/J.RSER.2015.09.098.
25. Huang X, Han S, Huang W, Liu X. Enhancing solar cell efficiency: the search for luminescent materials as spectral converters. Chem Soc Rev. 2012;42(1):173–201. https://doi.org/10.1039/ C2CS35288E.
26. Kumar KB, Gupta M, Mehta DS. Efficient sunlight harvesting with combined system of large Fresnel lens segmented mirror reflectors and compound parabolic concentrator without tracking sun for indoor daylight illumination. Renew Energy. 2023;202:1198–214. https://doi.org/10. 1016/J.RENENE.2022.11.117.
27. Jafari S, Byun YC. Prediction of the battery state using the digital twin framework based on the battery management system. IEEE Access. 2022;10:124685–96.
28. Jiang Y, Pu K. Multimodal biophotonics of semiconducting polymer nanoparticles. Acc Chem Res. 2018;51(8):1840–9. https://doi.org/10.1021/ACS.ACCOUNTS.8B00242/ASSET/ IMAGES/MEDIUM/AR-2018-002428_0007.GIF.
29. Joe HE, Yun H, Jo SH, Jun MBG, Min BK. A review on optical fiber sensors for environmental monitoring. Int J Precis Eng Manuf Green Technol. 2018;5(1):173–91. https://doi.org/10.1007/ S40684-018-0017-6.
30. Julus LJ, Roobert AA, Athanesious JJ. Role of photonics in energy crisis. 2023;205–222. https://doi.org/10.1007/978-981-99-2548-3_12.
31. Khonina SN, Kazanskiy NL, Butt MA. Optical fibre-based sensors—an assessment of current innovations. Biosensors 2023;13(9):835. https://doi.org/10.3390/BIOS13090835.
32. Kim BJ, Kim DH, Lee YY, Shin HW, Han GS, Hong JS, Mahmood K, Ahn TK, Joo YC, Hong KS, Park NG, Lee S, Jung HS. Highly efficient and bending durable perovskite solar cells: toward a wearable power source. Energy Environ Sci. 2015;8(3):916–21. https://doi.org/10. 1039/C4EE02441A.

33. Knebusch J, Gundlach J, Meier D, Meier T, Govers Y. Static testing using OFDR-based quasi continuous fiber optic strain sensing on the WiValdi wind turbine rotor blade. J Phys: Conf Ser. 2024;2767(4): 042005. https://doi.org/10.1088/1742-6596/2767/4/042005.
34. Lazaropoulos AG, Leligou HC. Integration of LiFi, BPL, and fiber optic technologies in smart grid backbone networks: a proposal for exploiting the LiFi LED street lighting networks of power utilities and smart cities. Sustainability 2024;16(2):686. https://doi.org/10.3390/SU16020686.
35. Lou D, Pang Q, Pei X, Dong S, Li S, Tan WQ, Ma L. Flexible wound healing system for pro-regeneration, temperature monitoring and infection early warning. Biosens Bioelectron. 2020;162: 112275. https://doi.org/10.1016/J.BIOS.2020.112275.
36. Märki I, Salt M, Herzig HP. Tunable microcavities in planar photonic crystals. 2006;6114:98–107. https://doi.org/10.1117/12.659901.
37. Marrero A, González J, Carta JA, Cabrera P. A new control algorithm to increase the stability of wind–hydro power plants in isolated systems: El Hierro as a case study. J Mar Sci Eng. 2023;11(223):335. https://doi.org/10.3390/JMSE11020335.
38. Massaro M, Secinaro S, Dal Mas F, Brescia V, Calandra D. Industry 4.0 and circular economy: an exploratory analysis of academic and practitioners' perspectives. Bus Strat Environ. 2021;30(2):1213–1231. https://doi.org/10.1002/BSE.2680.
39. Pelay U, Luo L, Fan Y, Stitou D, Rood M. Thermal energy storage systems for concentrated solar power plants. Renew Sustain Energy Rev. 2017;79:82–100. https://doi.org/10.1016/J.RSER.2017.03.139.
40. Pelucchi E, Fagas G, Aharonovich I, Englund D, Figueroa E, Gong Q, Hannes H, Liu J, Lu CY, Matsuda N, Pan JW. The potential and global outlook of integrated photonics for quantum technologies. Nat Rev Phys. 2022;4(3):194–208.
41. Mathias PC, Ganesh N, Cunningham BT. Application of photonic crystal enhanced fluorescence to a cytokine immunoassay. Anal Chem. 2008;80(23):9013-20. https://doi.org/10.1021/ac801377k
42. Measures RM. Laser remote sensing: fundamentals and applications. EOS Trans Am Geophys Union. 1985;66(40):686–686. https://doi.org/10.1029/EO066I040P00686-05.
43. Miller OD. Photonic design: From fundamental solar cell physics to computational inverse design. 2012.
44. Miles J. The application of laser anemometry to the measurement of airflow through a cascade of turbine blades. Optica Acta: Int J Optics. 1980;27(1):67–81. https://doi.org/10.1080/713820142.
45. Mohammadnia A, Ziapour MB, Ghaebi H, Khooban MH. Feasibility assessment of next-generation drones powering by laser-based wireless power transfer. Opt Laser Technol. 2021;143:107283. https://doi.org/10.1016/J.OPTLASTEC.2021.107283.
46. Ning K, Rubin J. Data driven artificial intelligence techniques in renewable energy system. 2021.
47. O'Brien JL, Furusawa A, Vučković J. Photonic quantum technologies. Nat Photon. 2009;3(12):687–695. https://doi.org/10.1038/nphoton.2009.229.
48. Ogunniyi EO, Pienaar HCVZ. Overview of battery energy storage system advancement for renewable (photovoltaic) energy applications. In: Proceedings of the 25th conference on the domestic use of energy, DUE 2017. pp. 233–239. https://doi.org/10.23919/DUE.2017.7931849.
49. Orr P, Booth C, Fusiek G, Niewczas P, Dysko A, Kawano F, Beaumont P. Distributed photonic instrumentation for smart grids. In: 2013 IEEE international workshop on applied measurements for power systems (AMPS). IEEE; 2013. pp. 63–67.
50. Shi Q, Dong B, He T, Sun Z, Zhu J, Zhang Z, Lee C. Progress in wearable electronics/photonics—moving toward the era of artificial intelligence and internet of things. InfoMat. 2020;2(6):1131–62. https://doi.org/10.1002/INF2.12122.
51. Singh V, Kumar R. Role of technological innovations in Industry 5.0. 73–98. https://doi.org/10.4018/979-8-3693-3550-5.CH006.
52. Peserico N, Ferreira de Lima T, Prucnal P, Sorger VJ. Emerging devices and packaging strategies for electronic-photonic AI accelerators: opinion. Opt Mater Express. 2022;12(4):1347–51.

53. Routray SK. The changing trends of optical communication. IEEE Potentials. 2014;33(1):28–33.
54. Xia F, Ang LY, Wang L, Vinel A. Internet of Things. Int J Commun Syst. 2012;25:1101–2. https://doi.org/10.1002/dac.2417.
55. Rekioua, D. Energy storage systems for photovoltaic and wind systems: a review. Energies. 2023;16(9):3893. https://doi.org/10.3390/EN16093893.
56. Ryoo G, Lee B, Shin S, Kim Y, Han JT, Jeong B, Park JH. Laser-assisted interfacial engineering for high-performance all-solid-state batteries. ChemElectroChem. 2023;10(20):202300349–202300350. https://doi.org/10.1002/celc.202300349.
57. Sangeethaa SN, Parthasarathi P, Jothimani S. Enabling technologies in IoT: energy, sensors, cloud computing, communication, integration, standards. Stud Syst Decis Control. 2024;472:493–511. https://doi.org/10.1007/978-3-031-29586-7_19.
58. Zheng D, Pauporté T, Schwob C, Coolen L. Models of light absorption enhancement in perovskite solar cells by plasmonic nanoparticles. Exploration. 2024;4(1):20220146. https://doi.org/10.1002/EXP.20220146.
59. Massaro A. Advanced electronic and optoelectronic sensors, applications, modelling and industry 5.0 perspectives. Appl Sci. 2023;13(7):4582.
60. Segun-Falade O, Osundare O, Kedi W. Developing innovative software solutions for effective energy management systems in industry (n.d.). Researchgate.Net. https://doi.org/10.51594/estj.v5i8.1517.
61. Shah R, Saha A, Najah Z, Prakash A, Reddy U, Nagpal A, Haripriya V. Role of quantum dots and nanostructures in photovoltaic energy conversion. E3S Web Conf. 2024;552:01096. https://doi.org/10.1051/e3sconf/202455201096.
62. Sharma S, Das S, Soni B, Sabir M. Review on the optics and photonics in environmental sustainability. 2024;359–385. https://doi.org/10.1007/978-3-031-64642-3_16.
63. Shekhar S, Bogaerts W, Chrostowski L, Bowers JE, Hochberg M, Soref R, Shastri BJ. Roadmapping the next generation of silicon photonics. Nat Commun. 2024;15(1):751.
64. Wu J, Lu Y, Feng S, Wu Z, Lin S, Hao Z, Yao T, Li X, Zhu H, Lin S. The interaction between quantum dots and graphene: the applications in graphene-based solar cells and photodetectors. Adv Funct Mater. 2018;28(50). https://doi.org/10.1002/ADFM.201804712.
65. Wang B, Zhou H, Li X, Yang G, Zheng P, Song C, Yuan Y, Wuest T, Yang H, Wang L. Human Digital Twin in the context of Industry 5.0. Robot Comp Integr Manuf. 2024;85. https://doi.org/10.1016/J.RCIM.2023.102626.
66. Wu Q, Tao M, Ng DW, Chen W, Schober R. Energy-efficient resource allocation for wireless powered communication networks. IEEE Trans Wirel Commun. 2015;15(3):2312-27.
67. Xu C, Lv M, Zhang K, Cao K, Wang G, Wei M, Peng B. Energy consumption and time-delay optimization of dependency-aware tasks offloading for industry 5.0 applications. IEEE Trans Cons Electron. 2024;70(1):1590–1600. https://doi.org/10.1109/TCE.2023.3338620.
68. Xu X, Lu Y, Vogel-Heuser B, Wang L. Industry 4.0 and Industry 5.0—inception, conception and perception. Journal of manufacturing systems. 2021;61:530–5.
69. Zhang W, Wang G, Yao G, Lu C, Liu Y. Study on fault monitoring technology of photovoltaic panel based on thermal infrared and optical remote sensing. Int Arch Photogramm Remote Sens Spatial Inf Sci. 2024;XLVIII-1–2024(1):855–60. https://doi.org/10.5194/ISPRS-ARCHIVES-XLVIII-1-2024-855-2024.
70. Zheng X, Zhang L. Photonic nanostructures for solar energy conversion. Energy Environ Sci. 2016;9(8):2511–32. https://doi.org/10.1039/C6EE01182A.
71. Zhironkin S, Abu-Abed F. Review of the transition to energy 5.0 in the context of non-renewable energy sustainable development. Energies. 2024;17(18):4723.
72. Zhu L, Akiyama H, Wang Y, Pan X. Theoretical modeling and ultra-thin design for multi-junction solar cells with a light-trapping front surface and its application to InGaP/GaAs/InGaAs 3-junction. Opt Express. 2022;30(20):35202–18. https://doi.org/10.1364/OE.466168.

# Chapter 10
# Challenges and Opportunities in Photonics Integration for Industry 5.0

**Tamanna Jindal, Sukhvir Singh, and Ranjana Jha**

**Abstract** Industry 5.0 marks a new epoch of humanitarian, intelligent automation, where the unified integration of modern technologies, including photonics, is key to revealing advanced industrial aptitudes. Photonics, which permits the transmission and manipulation of light, offers transformative assistances for industries, including earlier data communication, high accuracy sensing, and energy-effectual operations. The route to integrating photonics within the Industry 5.0 framework is fraught with challenges. These embrace complex technical barriers such as thermal management, photonic-electronic integration, scalability matters in material limitations, and the need for greater standardization. Moreover, a lack of skilled labor and expertise in photonic technologies hampers widespread adoption. Progressions in photonics are lashing innovation in key Industry 5.0 areas, from autonomous systems and virtual reality, to real-time monitoring in smart manufacturing and energy-competent telecommunications. Developing technologies such as quantum photonics, AI-improved optical systems, and flexible photonic devices are poised to reshape industries by enhancing human–machine collaboration and promoting sustainability. The chapter delves both the challenges and possibilities of photonics integration, even insights into ongoing research and the strategic importance of potential of photonics in shaping Industry 5.0.

**Keywords** Photonics · Industry 5.0 · AI-enhanced optical systems · Sustainable technologies · Data communication

## 10.1 Introduction

The Industry 5.0 represents a transformative leap from the automation-driven, data-centric model of Industry 4.0 to a new criterion where human inventiveness and collaboration with modern technologies play a central role [1]. While Industry 4.0

T. Jindal (✉) · S. Singh · R. Jha
Research Lab for Energy System, Department of Physics, Netaji Subhas University of Technology, New Delhi, India
e-mail: tamannajindal3@gmail.com



V. Bhardwaj et al. (eds.), *Photonics and Optoelectronics in Industry 5.0*, Advances in Optics and Optoelectronics, https://doi.org/10.1007/978-981-96-7594-4_10

was principally concerned on automation, digitization, and the interconnectivity of systems and machines, Industry 5.0 accentuate the flawless integration of human intelligence with cutting-edge technologies to create more flexible, and personalized production systems [2, 3]. This shift is not only about enhancing productivity but also about making industries more sustainable, resilient, and humanitarian [4].

At this new industrial uprising, the science and technology of light is preferably known as photonics. Photonics has grown to be an essential component for next-generation industrial systems, playing a vital role in different applications such as high-speed data communication, precision sensing [5, 6], energy-efficacious computing [7], and high-tech manufacturing processes [8, 9]. The high data transfer rates given by the optical communication system [10] are vital for dealing with the massive amounts of data generated by current industrial systems [11]. The integration of photonics with other evolving technologies, such as Internet of Things (IoT), artificial intelligence and quantum computing, is anticipated to drive profound variations across industries, from health care and telecommunications to aerospace and automotive. Though, the integration of photonics into Industry 5.0 also represents noteworthy challenges [12]. Technical barriers, such as complexities of photonic-electronic co-integration, thermal management and miniaturization, need to be overcome for photonics to reach its full potential. Furthermore, matters related to manufacturing scalability, material confines, and the lack of standardization across platforms are grave sprints that must be addressed. Contempt these challenges, the opportunities offered by photonics are vast, as innovations in optical technologies continue to hasten.

This chapter explores the challenges and opportunities of photonics integration in the framework of Industry 5.0. Investigating the practical, economic, and societal influences of photonics on next-generation industries, we will explore how photonic technologies are poised to play a transformative role in shaping the future of industrial ecosystems. Moving ahead where human–machine association and sustainable manufacturing are dominant goals, photonics will be instrumental in permitting the vision of Industry 5.0.

## 10.2 Challenges in Photonics Integration for Industry 5.0

### *10.2.1 Technical Barriers*

Though photonics suggestions transformative potential for Industry 5.0, its integration into industrial systems poses some significant technical challenges [13]. These barriers must be addressed to fully connect the capabilities of photonics in smart, human-centric industries. Below are roughly of the key technical barriers in photonics integration:

- Miniaturization and Integration Complexities

The main challenges in photonics integration are the necessity for miniaturization [14], particularly when integrating photonic apparatuses into increasingly compact devices. Photonic systems, including, modulators, detectors, and lasers [15–17], are typically larger than their electronic complements. Furthermore, combining photonic devices with electronic circuits involves precise alignment and sophisticated packaging techniques to ensure high performance and reliability. The co-integration of photonics and electronics on a single chip, recognized as photonic-electronic integration, is a critical technical hurdle that must be overcome to attain more competent systems for Industry 5.0 applications.

- Thermal Management and Heat Dissipation

In photonic systems, handling heat is additional substantial technical barrier. Though photonics-based communication systems are usually more energy-saving than their electronic complements, they still produce heat, chiefly in high-power applications such as amplifiers and lasers. Effective thermal management is indispensable to prevent performance signal integrity, degradation and ensure the long-term consistency of photonic components. Still, integrating active cooling mechanisms within the limitations of miniaturized systems remains a challenge. Advanced cooling technologies and innovative material designs are desired to address these thermal management issues in high-performance photonic systems.

- Material Limitations

Silicon photonics has seemed as a foremost platform due to its compatibility with current semiconductor fabrication measures, but silicon has intrinsic limitations, predominantly in its incapability to efficiently emit light [18]. This limitation necessitates the use of substitute materials, such as compound semiconductors [e.g., indium phosphide and gallium arsenide] or emerging materials like graphene and perovskites are more competent for certain photonic applications [19]. Nevertheless, incorporating these materials into mainstream manufacture procedures while conserving compatibility with existing technologies remains a multifaceted challenge.

- Testing and Quality Assurance

Photonic arrangements require laborious testing to confirm performance and safety, particularly in grave industrial applications. Though, testing photonic devices is more intricate than testing traditional electronic systems. Photonic mechanisms operate at extremely high speeds and interact with light in ways that are difficult to measure using standard electronic testing equipment. Emerging specialized testing procedures and tools for photonic systems is essential to guarantee the quality of these systems, but the current lack of consistent testing practices remains a barrier.

### *10.2.2 Manufacturing and Scalability Issues*

The potential to convert industries by permitting ultra-fast communication, accurate sensing, and energy-effectual systems. Still, achieving widespread adoption of photonic technologies is hindered by significant manufacturing and scalability challenges [20]. The subsequent are key issues related to these obstacles in photonics integration:

- High Manufacturing Costs

Photonic components comprise resource-rigorous processes such as modulators, lasers, waveguides, and photonic integrated circuits [PICs]. These measures often necessitate specialized materials [e.g., compound semiconductors like indium phosphide], progressive fabrication facilities, and exact alignment techniques, all of which subsidize to high production costs.

Basically, in electronics, where silicon-based manufacturing is well-established and highly scalable, however the production of photonic components often involves smaller production volumes, leading to less cost-efficiency [21]. Economies of scale are difficult to achieve due to the relatively low demand compared to traditional semiconductor devices, making photonics less cost-competitive in large-scale industrial applications.

- Lack of Standardized Manufacturing Processes

Photonic devices frequently vary based on the specific materials and fabrication techniques used, which hinders mass production and limits cross-platform compatibility [22]. The primary challenges in scaling up photonics integration is the lack of standardized manufacturing processes across the industry.

Manufacturing of photonics is still highly fragmented, contrasting the semiconductor industry, where standardized procedures like Complementary Metal–Oxide–Semiconductor (CMOS) technology have enabled large-scale integration and cost reduction. The absence of a unified, scalable platform for producing photonic integrated circuits (PICs) complicates efforts to develop cost-effective, high-volume manufacturing.

- Material Challenges for Scalable Production

Intended for ascendable production gallium arsenide, indium phosphide, or incipient materials like graphene and perovskites are frequently used in photonics devices. These materials require particular fabrication techniques that are not always compatible with large-scale production. Unlike silicon, materials are more expensive and difficult to exertion [23].

Furthermore, the multiple materials with different properties are combined on the same chip also called the assorted integration of materials, (e.g., silicon photonics combined with III–V semiconductors for light emission), which introduces further challenges in terms of fabrication complexity and cost. This adds to the difficulty of achieving scalable production for photonic devices.

- Packaging and Assembly

Packaging photonic devices is more complex and costly than packaging electronic components due to the need for highly precise optical alignment and coupling of different components (e.g., lasers, detectors, and modulators). Optical packaging often involves intricate assembly steps, such as aligning fibers with waveguides, which are labor-intensive and not easily automated [24].

Environmental factors such as temperature, humidity, and vibration, photonic devices are sensitive to it and making protective packaging critical to ensure their long-term reliability. Developing cost-effective, scalable packaging solutions that maintain the performance of photonic components is a major challenge for manufacturers.

### *10.2.3 Interoperability and Standardization*

The potential to revolutionize industries by enabling faster, more efficient, and intelligent systems. Substantial challenges related to interoperability and standardization hinder the widespread adoption of photonics in industrial ecosystems [25]. These issues must be addressed to ensure that photonic technologies can seamlessly integrate with existing systems, communicate across platforms, and be produced at scale with uniformity and reliability. Below are the key interoperability and standardization issues in photonics integration for Industry 5.0:

- Lack of Unified Standards

Unlike the semiconductor and electronics industries, which have well-established standards (e.g., CMOS for integrated circuits), photonics lacks a unified, globally accepted set of standards for manufacturing, design, and testing. This absence of standardization makes it difficult for different photonic components, devices, and systems to interoperate smoothly. Different manufacturers may use varying processes, materials, and architectures, leading to compatibility issues when integrating photonic components from multiple suppliers [26]. Without common standards, the design and production of photonic systems become fragmented, limiting the ability to achieve large-scale adoption and interoperability across industries.

- Interoperability Between Photonic and Electronic Systems

Industry 5.0 envisions systems where photonics and electronics work in tandem, combining the high-speed data transmission and sensing capabilities of photonics with the processing power and control functions of electronics. However, integrating these two domains presents significant interoperability challenges. Photonic and electronic components often operate using different signal formats, data transmission protocols, and power requirements [27]. Ensuring seamless communication between photonic and electronic systems requires specialized interface technologies that can convert signals between light and electricity without introducing latency or signal degradation. The absence of standardized protocols and interfaces for photonic-electronic co-integration complicates the design and deployment of hybrid systems, which are essential for Industry 5.0 applications such as real-time monitoring, autonomous systems, and AI-driven manufacturing.

- Compatibility with Legacy Systems

Many industrial environments are built on legacy electronic systems that have been optimized over decades for specific applications. Introducing photonics into these environments requires compatibility with existing communication networks, control systems, and data infrastructure. Interfacing photonic technologies with older, established systems can be challenging due to differences in data formats, signal processing methods, and physical architectures. Without standardized interoperability solutions, the integration of photonics into legacy industrial systems becomes costly and complex, limiting the potential benefits of photonic technologies.

### *10.2.4 Reliability and Durability*

In the context of Industry 5.0, the reliability and durability of photonic devices are critical factors that determine their effectiveness and longevity in industrial applications. As industries increasingly integrate photonic technologies into their operations, ensuring the long-term performance and environmental stability of these devices becomes essential [24]. Below are the key considerations related to the reliability and durability of photonic systems:

- Performance Consistency: Photonic devices, such as lasers, sensors, and optical components, must maintain consistent performance over extended periods. Factors such as degradation of materials, alignment issues, and optical losses can adversely affect the reliability of these devices [28]. Manufacturers must adopt robust design and manufacturing practices to enhance the longevity of photonic components, including the use of high-quality materials and precision engineering techniques.
- Failure Modes: Understanding the potential failure modes of photonic devices is crucial for improving reliability. Common failure mechanisms include thermal degradation, mechanical stress, and optical interference. Conducting thorough reliability testing and implementing predictive maintenance strategies can help identify and mitigate these risks before they lead to operational failures.
- Redundancy and Fault Tolerance: Incorporating redundancy into photonic systems can enhance reliability. By designing systems with backup components or alternative pathways for data transmission, industries can ensure continuous operation even in the event of component failure [29]. Implementing fault-tolerant designs can significantly reduce downtime and increase overall system reliability.
- Resistance to Temperature Variations: Photonic devices often operate in environments with fluctuating temperatures. Ensuring that these devices can function

effectively across a wide temperature range is critical. Materials used in photonic systems must have low thermal expansion coefficients and high thermal stability to prevent performance degradation due to temperature variations.

- Moisture Resistance: Exposure to moisture can adversely affect the performance and reliability of photonic devices. Components such as sensors and optical fibers must be designed to withstand humidity and moisture ingress [30]. Utilizing protective coatings, hermetic packaging, and moisture-resistant materials can enhance the durability of photonic systems in humid environments.
- Mechanical Stability: Photonic devices may also be exposed to vibrations and mechanical stresses in industrial settings. Ensuring that these devices are mechanically robust and capable of withstanding harsh conditions is essential. This includes designing for shock resistance and employing materials that can tolerate mechanical wear over time.
- Chemical Resistance: In many industrial applications, photonic devices may come into contact with chemicals, solvents, and other corrosive substances. The materials used in photonic systems must be chemically resistant to prevent degradation and maintain performance in such environments.

### *10.2.5 Workforce and Expertise Gaps*

As industries transition toward integrating advanced technologies within the framework of Industry 5.0, a critical challenge arises: the gaps in workforce skills and expertise in the field of photonics. For the successful adoption and implementation of photonic technologies, a skilled workforce with specialized training and knowledge is essential. Below are the key facets of these gaps that need to be addressed to facilitate the effective integration of photonics into contemporary industrial settings.

- Need for Specialized Training in Photonics

Many existing educational programs in engineering and technology do not adequately cover photonics. Developing robust curricula that integrate photonics into the training of future engineers and technicians is essential. This includes hands-on training, laboratory work, and real-world applications that provide students with practical experience in photonics. As photonics technology continues to evolve rapidly, ongoing professional development is crucial for existing professionals in the field. Training programs, workshops, and certification courses can help upskill the workforce and keep them abreast of the latest advancements in photonic technologies. Encouraging a culture of lifelong learning is vital to bridging the expertise gap.

- Limited Industry Knowledge and Skilled Labor Shortage

Many companies may lack awareness of the potential benefits that photonics can bring to their operations. This limited understanding can hinder the investment in and adoption of photonic technologies [30]. Raising awareness through industry workshops, conferences, and collaborative research initiatives can help disseminate knowledge about the advantages and applications of photonics.

The current labor market faces a shortage of skilled workers with expertise in photonics. This shortage can slow the pace of innovation and implementation in industries that could benefit from photonics integration [29]. Companies need to actively engage in workforce development initiatives, such as partnerships with educational institutions, internships, and apprenticeship programs, to cultivate the next generation of photonics professionals.

## 10.3 Opportunities in Photonics Integration

The incorporation of photonics across different industries offers numerous opportunities that can greatly improve productivity, efficiency, and innovation. As technology evolves, the range of applications for photonics broadens, leading to transformative changes in various sectors. Here are some of the key opportunities that emerge from advancements in photonic technologies.

- Advancements in Photonic Technologies

Advancements in photonic technologies have led to the development of highly sensitive sensors for a range of applications. Photonic sensors can detect variations in temperature, pressure, and environmental conditions with exceptional accuracy, making them essential in industries like health care, manufacturing, and environmental monitoring [28]. For instance, optical sensors employed in medical diagnostics allow for real time, non-invasive health monitoring of patients, facilitating more timely and precise medical interventions. The potential of photonics to improve sensing and measurement capabilities paves the way for innovation and quality enhancement across various sectors.

- Automation and Smart Manufacturing

The integration of photonics into manufacturing processes can enhance automation and the development of smart factories. Photonic technologies enable precise control and monitoring of production systems, allowing for real-time data analysis and decision-making. For instance, machine vision systems using photonic sensors can identify defects in products during the manufacturing process, reducing waste and improving quality. Additionally, the combination of photonics with artificial intelligence can lead to advanced predictive maintenance solutions, minimizing downtime and maximizing productivity.

- Innovations in Consumer Electronics

Evolution in photonics is fostering innovation in consumer electronics, with applications that span from augmented reality (AR) and virtual reality (VR) to high-resolution displays [31]. These photonic technologies enhance user experiences by creating more immersive and interactive environments. For instance, photonic integrated circuits can facilitate the development of compact and efficient optical devices for AR glasses, thereby enhancing the functionality and portability of consumer electronics. As photonic technologies continue to evolve, we can expect the emergence of new products and services that address the changing needs of consumers.

- Enhanced Data Transmission and Communication

Photonics offers a pathway to superior data transmission capabilities compared to traditional electronic methods. Fiber optic communication systems, for instance, can transmit data over long distances with minimal loss and at incredibly high speeds [32]. This advancement supports the growing demand for bandwidth in industries such as telecommunications, where the need for fast, reliable communication networks is paramount. The integration of photonics can lead to the development of more efficient data centers and improved connectivity, facilitating seamless communication in an increasingly digital world.

- Photonics in Green Manufacturing

The adoption of photonics in green manufacturing represents a significant advancement in sustainable industrial practices. By leveraging photonic technologies, industries can enhance their operational efficiency while minimizing environmental impact.

Photonics can enhance environmental monitoring capabilities within manufacturing environments. Optical sensors can be employed to detect pollutants, measure emissions, and monitor environmental conditions in real time [33]. This data enables manufacturers to ensure compliance with environmental regulations and standards, facilitating proactive measures to reduce their ecological impact. By adopting photonic technologies for environmental monitoring, companies can maintain transparency and demonstrate their commitment to sustainability.

## 10.4 Emerging Solutions and Research Directions

The incorporation of photonics into Industry 5.0 offers significant opportunities for innovation and enhanced efficiency. As industries strive to leverage advanced photonic technologies, the development of emerging solutions and research initiatives is becoming increasingly important. Here are some key areas where new solutions are being created and research efforts are concentrated to promote the integration of photonics across various sectors.

- Integrated Photonics

The trend toward integrated photonics is gaining momentum, aiming to miniaturize and enhance the functionality of photonic devices by combining multiple functions onto a single chip. This integration can lead to smaller, more efficient systems that consume less power and are easier to manufacture [34]. Research efforts are focused on developing photonic integrated circuits (PICs) that can perform complex tasks, such as data processing, sensing, and communication, all within a compact footprint. This advancement has the potential to revolutionize telecommunications, healthcare diagnostics, and consumer electronics.

- Machine Learning and AI in Photonics

The intersection of machine learning, artificial intelligence (AI), and photonics is a promising area for research and development. By leveraging AI algorithms, photonic systems can be optimized for real-time data analysis, predictive maintenance, and adaptive manufacturing processes [35]. For example, AI-driven algorithms can analyze data from photonic sensors to identify patterns and anomalies, enabling more efficient operation and maintenance of manufacturing equipment. This integration can lead to smarter, more autonomous industrial environments.

- Quantum Photonics

Quantum photonics represents a groundbreaking area of research with the potential to transform industries. By utilizing the principles of quantum mechanics, researchers are exploring photonic technologies that can enable ultra-secure communication, advanced quantum computing, and enhanced sensing capabilities [36]. Figure 10.1 outlines the three key elements of integrated quantum communication [37, 38]. Developments in quantum key distribution (QKD) and quantum-enhanced imaging are paving the way for secure data transmission and high-precision measurements [39]. This area of research holds promise for applications in cybersecurity, telecommunications, and medical diagnostics.

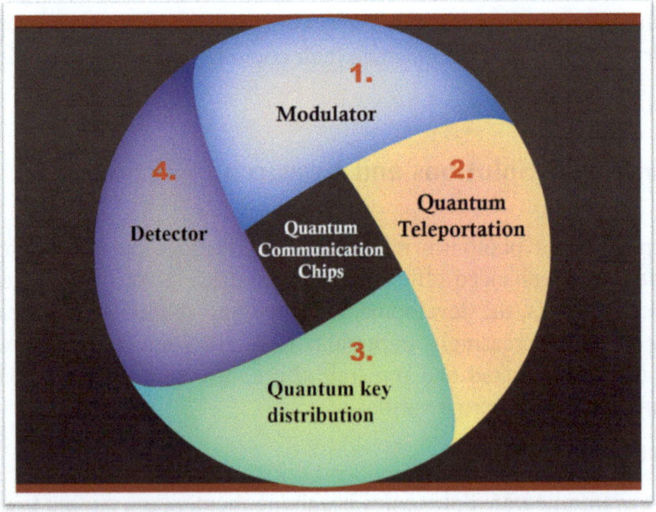


**Fig. 10.1** Quantum communication chips [37, 38]

- Sustainable Photonic Solutions

As industries increasingly prioritize sustainability, research is being directed toward developing environmentally friendly photonic technologies. This includes improving the efficiency of solar cells, creating biodegradable photonic materials, and developing photonic systems that minimize energy consumption [32]. Innovations in sustainable photonics can significantly contribute to the reduction of carbon footprints in manufacturing and energy generation, aligning with the broader goals of Industry 5.0.

## 10.5 Conclusion

In this chapter, we explored the integration of photonics within Industry 5.0 presents a landscape rich with both challenges and opportunities. As industries adapt to the demands of a rapidly evolving technological environment, the potential for photonics to revolutionize processes and enhance efficiency is undeniable. However, overcoming obstacles such as manufacturing complexities, interoperability issues, and the need for skilled personnel will be crucial for realizing these advancements. Looking ahead, photonics is poised to become a key enabler for Industry 5.0, facilitating the evolution of manufacturing and industrial practices into a more interconnected, efficient, and sustainable framework. As research continues to advance in areas such as integrated photonics, quantum technologies, and machine learning applications, we can expect to see innovative solutions that address current challenges and unlock new possibilities.

The focus on sustainability and energy efficiency will further drive the adoption of photonic technologies as industries strive to meet regulatory demands and consumer expectations for environmentally responsible practices. As stakeholders across sectors collaborate to establish standards, invest in workforce development, and explore emerging applications, the role of photonics in shaping the future of industry will be increasingly significant.

In conclusion, while challenges remain, the potential for photonics to revolutionize industries within the context of Industry 5.0 is undeniable. By addressing these challenges head-on and leveraging the opportunities presented by photonics, industries can embark on a path toward a smarter, more sustainable future.

## References

1. Massaro A. Advanced electronic and optoelectronic sensors, applications, modelling and industry 5.0 perspectives. Appl Sci. 2023;13(7). https://www.mdpi.com/2076-3417/13/7/4582.
2. Yang J, Liu Y, Morgan PL. Human–machine interaction towards Industry 5.0: Human-centric smart manufacturing. Digit Eng. 2024;2:100013. https://www.sciencedirect.com/science/article/pii/S2950550X2400013X.

3. Leng J, Sha W, Wang B, Zheng P, Zhuang C, Liu Q, et al. Industry 5.0: prospect and retrospect. J Manuf Syst. 2022;65:279–95. https://www.sciencedirect.com/science/article/pii/S0278612522001662.
4. Rožanec JM, Novalija I, Zajec P, Kenda K, Tavakoli Ghinani H, Suh S, et al. Human-centric artificial intelligence architecture for industry 5.0 applications. Int J Prod Res. 2023;61(20):6847–72. https://doi.org/10.1080/00207543.2022.2138611.
5. Kazanskiy NL, Khonina SN, Butt MA. Advancement in silicon integrated photonics technologies for sensing applications in near-infrared and mid-infrared region: a review. Photonics. 2022;9(5). https://www.mdpi.com/2304-6732/9/5/331.
6. Dhote C, Singh A, Kumar S. Silicon photonics sensors for biophotonic applications—a review. IEEE Sens J. 2022;22(19):18228–39.
7. Dhasarathan V, Sahu SK, Nguyen TK, Palai G. Realization of all logic gates using metamaterials based three dimensional photonics structures: a future application of 3D photonics to optical computing. Optik [Stuttg]. 2020;202:163723. https://www.sciencedirect.com/science/article/pii/S0030402619316213.
8. Maddikunta PKR, Pham QV, B P, Deepa N, Dev K, Gadekallu TR, et al. Industry 5.0: a survey on enabling technologies and potential applications. J Ind Inf Integr. 2022;26:100257. https://www.sciencedirect.com/science/article/pii/S2452414X21000558.
9. Pelucchi E, Fagas G, Aharonovich I, Englund D, Figueroa E, Gong Q, et al. The potential and global outlook of integrated photonics for quantum technologies. Nat Rev Phys. 2022;4(3):194–208. https://doi.org/10.1038/s42254-021-00398-z.
10. Meng Y, Chen Y, Lu L, Ding Y, Cusano A, Fan JA, et al. Optical meta-waveguides for integrated photonics and beyond. Light Sci Appl. 2021;10(1):235. https://doi.org/10.1038/s41377-021-00655-x.
11. Butt MA, Kazanskiy NL, Khonina SN, Voronkov GS, Grakhova EP, Kutluyarov R V. A review on photonic sensing technologies: status and outlook. Biosensors [Basel]. 2023;13(5). https://www.mdpi.com/2079-6374/13/5/568.
12. Amiri IS, Azzuhri SR Bin, Jalil MA, Hairi HM, Ali J, Bunruangses M, et al. Introduction to photonics: principles and the most recent applications of microstructures. Micromachines [Basel]. 2018;9(9). https://www.mdpi.com/2072-666X/9/9/452.
13. Polymeni S, Plastras S, Skoutas DN, Kormentzas G, Skianis C. The impact of 6G-IoT technologies on the development of agriculture 5.0: a review. Electronics [Basel]. 2023;12(12). https://www.mdpi.com/2079-9292/12/12/2651.
14. Carrasco-Casado A, Shiratama K, Kolev D, Ono F, Tsuji H, Toyoshima M. Miniaturized multi-platform free-space laser-communication terminals for beyond-5G networks and space applications. Photonics. 2024;11(6). https://www.mdpi.com/2304-6732/11/6/545.
15. Tang M, Park JS, Wang Z, Chen S, Jurczak P, Seeds A, et al. Integration of III-V lasers on Si for Si photonics. Prog Quantum Electron. 2019;66:1–18. https://www.sciencedirect.com/science/article/pii/S0079672719300163.
16. Witzens J. High-speed silicon photonics modulators. Proc IEEE. 2018;106(12):2158–82.
17. Gyger S, Zichi J, Schweickert L, Elshaari AW, Steinhauer S, Covre da Silva SF, et al. Reconfigurable photonics with on-chip single-photon detectors. Nat Commun. 2021;12(1):1408. https://doi.org/10.1038/s41467-021-21624-3.
18. Wu J, Xu X, Nguyen TG, Chu ST, Little BE, Morandotti R, et al. RF photonics: an optical microcombs' perspective. IEEE J Sel Top Quantum Electron. 2018;24(4):1–20.
19. Piggott AY, Ma EY, Su L, Ahn GH, Sapra NV, Vercruysse D, et al. Inverse-designed photonics for semiconductor foundries. ACS Photonics. 2020;7(3):569–75. https://doi.org/10.1021/acsphotonics.9b01540.
20. Tzampazaki M, Zografos C, Vrochidou E, Papakostas GA. Machine vision—moving from industry 4.0 to industry 5.0. Appl Sci. 2024;14(4). https://www.mdpi.com/2076-3417/14/4/1471.
21. Fernández-Miguel A, García-Muiña FE, Jiménez-Calzado M, Melara San Román P, Fernández del Hoyo AP, Settembre-Blundo D. Boosting business agility with additive digital molding: an Industry 5.0 approach to sustainable supply chains. Comput Ind Eng. 2024;192:110222. https://www.sciencedirect.com/science/article/pii/S0360835224003437.

22. Chi HR, de Fátima Domingues M, Zhu H, Li C, Kojima K, Radwan A. Healthcare 5.0: in the perspective of consumer internet-of-things-based fog/cloud computing. IEEE Trans Cons Electron. 2023;69(4):745–55.
23. Donmezer S, Demircioglu P, et al. Revolutionizing industry 5.0: harnessing the power of digital human modelling. In: Durakbasa Numan M. and Gençylmaz MG, editor. Industrial engineering in the industry 40 era. Cham: Springer Nature Switzerland; 2024. p. 223–35.
24. Elim HI, Zhai G. Control system of multitasking interactions between society 5.0 and industry 5.0: a conceptual introduction & its applications. J Phys Conf Ser. 2020;1463(1):12035. https://doi.org/10.1088/1742-6596/1463/1/012035.
25. Espina-Romero L, Guerrero-Alcedo J, Goñi Avila N, Noroño Sánchez JG, Gutiérrez Hurtado H, Quiñones Li A. Industry 5.0: tracking scientific activity on the most influential industries, associated topics, and future research agenda. Sustainability. 2023;15(6). https://www.mdpi.com/2071-1050/15/6/5554.
26. Rahim A, Goyvaerts J, Szelag B, Fedeli JM, Absil P, Aalto T, et al. Open-access silicon photonics platforms in Europe. IEEE J Sel Top Quantum Electron. 2019;25(5):1–18.
27. Soref R. Emerging SiGeSn integrated-photonics technology. In: 2016 IEEE photonics society summer topical meeting series [SUM]. 2016. p. 100–1.
28. Huang C, Zhang H, Sun H. Ultraviolet optoelectronic devices based on AlGaN-SiC platform: Towards monolithic photonics integration system. Nano Energy. 2020;77:105149. https://www.sciencedirect.com/science/article/pii/S2211285520307278.
29. You J, Luo Y, Yang J, Zhang J, Yin K, Wei K, et al. Hybrid/integrated silicon photonics based on 2d materials in optical communication nanosystems. Laser Photon Rev. 2020;14(12):2000239. https://doi.org/10.1002/lpor.202000239.
30. Cheng Z, Cao R, Wei K, Yao Y, Liu X, Kang J, et al. 2D materials enabled next-generation integrated optoelectronics: from fabrication to applications. Adv Sci. 2021;8(11):2003834. https://doi.org/10.1002/advs.202003834.
31. Camposeo A, Persano L, Farsari M, Pisignano D. Additive manufacturing: applications and directions in photonics and optoelectronics. Adv Opt Mater. 2019;7(1):1800419. https://doi.org/10.1002/adom.201800419.
32. Shi Y, Zhang Y, Wan Y, Yu Y, Zhang Y, Hu X, et al. Silicon photonics for high-capacity data communications. Photon Res. 2022;10(9):A106–A134. https://opg.optica.org/prj/abstract.cfm?URI=prj-10-9-A106.
33. Sridarshini T, Chidambaram P, et al. Current and future horizon of optics and photonics in environmental sustainability. Sustain Comput: Inf Syst. 2022;36:100815. https://www.sciencedirect.com/science/article/pii/S2210537922001469.
34. Barwicz T, Peng B, Leidy R, Janta-Polczynski A, Houghton T, Khater M, et al. Integrated metamaterial interfaces for self-aligned fiber-to-chip coupling in volume manufacturing. IEEE J Sel Top Quantum Electron. 2019;25(3):1–13.
35. Liu Z, Zhu D, Raju L, Cai W. Tackling photonic inverse design with machine learning. Adv Sci. 2021;8(5):2002923. https://doi.org/10.1002/advs.202002923.
36. Turunen M, Brotons-Gisbert M, Dai Y, Wang Y, Scerri E, Bonato C, et al. Quantum photonics with layered 2D materials. Nat Rev Phys. 2022;4(4):219–36. https://doi.org/10.1038/s42254-021-00408-0.
37. Luo W, Cao L, Shi Y, Wan L, Zhang H, Li S, et al. Recent progress in quantum photonic chips for quantum communication and internet. Light Sci Appl. 2023;12(1):175. https://doi.org/10.1038/s41377-023-01173-8.
38. Siew SY, Li B, Gao F, Zheng HY, Zhang W, Guo P, et al. Review of silicon photonics technology and platform development. J Lightwave Technol. 2021;39(13):4374–89. https://opg.optica.org/jlt/abstract.cfm?URI=jlt-39-13-4374.
39. Solntsev AS, Agarwal GS, Kivshar YS. Metasurfaces for quantum photonics. Nat Photonics. 2021;15(5):327–36. https://doi.org/10.1038/s41566-021-00793-z.

# Chapter 11
# Emerging Trends and Future Directions in Photonics and Optoelectronics for Industry 5.0

**Garima Chanana**

**Abstract** The rapid evolution of Industry 5.0 emphasizes human-centric, sustainable, and intelligent systems, where photonics and optoelectronics play a pivotal role in shaping the future of industrial applications. This chapter explores the fundamental principles of photonics, key technological advancements, and their integration into next-generation industrial systems. Emerging trends such as smart photonic sensors, integrated photonics, quantum photonics, AI-driven photonic design, and energy-efficient optoelectronic technologies are analysed for their transformative potential. Furthermore, the chapter discusses applications in smart manufacturing, high-speed communication, health care, security, and defence, highlighting how photonics enhances efficiency, precision, and sustainability. While the field presents challenges related to scalability, cost, and hybrid integration, ongoing research is driving innovative solutions to overcome these barriers. Future directions, including advancements in quantum photonics, AI-optimized optical systems, and neuromorphic photonics, are expected to redefine industrial automation and communication. Ultimately, photonics and optoelectronics will serve as fundamental enablers of Industry 5.0, fostering a more connected, intelligent, and sustainable industrial landscape.

**Keywords** Photonics · Optoelectronics · Industry 5.0 · Quantum photonics · Smart manufacturing · Integrated photonics

G. Chanana (✉)
School of Engineering and Technology, Vivekananda Institute of Professional Studies, Technical Campus, Delhi, India
e-mail: garima.chanana@vips.edu

 
V. Bhardwaj et al. (eds.), *Photonics and Optoelectronics in Industry 5.0*, Advances in Optics and Optoelectronics, https://doi.org/10.1007/978-981-96-7594-4_11

## 11.1 Introduction

Industry 5.0 [1] has introduced a revolutionary change in industrial growth. It seeks to go beyond automation-centric emphasis of Industry 4.0 to a human-oriented, sustainable, and resilient strategy. Whereas Industry 4.0 was mostly about digital transformation through Internet of Things, artificial intelligence, big data, and cyber-physical systems. Industry 5.0 seeks to combine and balance new age technologies with human innovation and environmental management [2]. This transition has necessitated demands for more efficient, adaptive, and sustainable industrial solutions that can embrace existing leading technologies as well as human collaboration with ethical code.

Photonics and optoelectronics [3] are coming forward as integral technology for becoming the basis for practical realization of Industry 5.0. In the present scenario, industries are going to develop an ultra-fast, energy-efficient, and accurate systems. Photonics [4] which is the study of light generation, manipulation, and detection, enables numerous advanced applications, such as laser-based manufacturing, fibre-optic communication, optical sensors, and imaging systems. Likewise, optoelectronics [5], which is concerned with electronic devices that respond to light, encompasses technologies like light-emitting diodes (LEDs), photodetectors, and photovoltaic systems, which makes a valuable contribution to energy efficiency, high-speed data transmission, and sophisticated sensing applications.

The shift from Industry 4.0 to Industry 5.0 is characterized by moving away from totally automated and data-centric processes to an environment that focuses on human–machine collaboration [6], sustainability, and resilience [7]. Photonics and optoelectronics in this context are poised to tackle future challenges in numerous fields. For example, photonic sensors [8] allow rapid and precise measurements in a variety of applications in health care, environmental monitoring, defence, and security with applications in medical imaging, real-time pollutant sensing, and laser-based surveillance systems. Optical computing [9, 10] and quantum photonics [11] enable ultra-fast data processing, supporting the next generation of computing and artificial intelligence systems. Optoelectronic devices [12] are transforming diagnostics, imaging, and targeted therapy in the medical field by allowing minimally invasive procedures.

Another major influencer of Industry 5.0 is the focus on energy-efficient and sustainable solutions. Photonic technologies, including silicon photonics [13, 14] and energy-efficient light-emitting diodes [15] provide great advantage in minimizing power consumption in industries. Green photonics [16, 17], which emphasizes environmentally friendly optical technologies, aims to maximize energy efficiency and reduce the carbon footprint of industrial processes. Solar photovoltaics [18–20] is an important subfield of optoelectronics which is increasingly finding applications in industries transition to renewable energy sources to address global sustainability objectives.

In addition, the development of human-centric industrial systems in Industry 5.0 [21, 22] requires adaptive and intelligent optoelectronic solutions. Virtual and augmented reality systems [23], driven by optical displays and sensors, are facilitating human-industrial system interaction, safety, training, and efficiency.

This chapter delves into the revolutionary role of photonics and optoelectronics in defining Industry 5.0. It discusses the technological developments, applications, challenges, and future trends in this field. Through a detailed analysis of recent advances in optical sensing, quantum photonics, green optoelectronics, and smart light-based systems, this analysis seeks to give insights into how these technologies will drive revolution in Industry 5.0. In the end, photonic and optoelectronic solutions will need to be combined in order to create a more intelligent, cleaner, and more sustainable industrial ecosystem for Industry 5.0.

## 11.2 Fundamentals of Photonics and Optoelectronics

Photonics and optoelectronics [24] are fundamental domains that have had a major impact on contemporary technological developments. Photonics refers to the science of generating, and controlling light-based phenomena, while optoelectronics is the study of the interaction between light and electronic devices. These technologies are central to many innovations in biomedical science [25], additive manufacturing [26], and sustainable photonics [27], and they are vital in determining the future of industrial applications under Industry 5.0.

At the core of photonics is the basic behaviour of light, which can be explained by principles like reflection, refraction, diffraction, and interference [28]. The capability to control light at different wavelengths allows for a variety of functionalities, ranging from high-speed data transmission [29] to precision manufacturing [30].

It is vital to explore the main mechanisms of photonics to utilize its power for maximum utility for various applications. They involve fundamental light-matter interactions, wave-optical and quantum-optical effects, nonlinear effects, and optical device level processes as illustrated in Fig. 11.1.

Fundamental light-matter interaction processes [31] determine the behaviour of light when it passes through various media. Reflection is the bouncing of light off a surface according to the law of reflection, while refraction involves bending of light as it crosses material boundaries of different refractive indices. Diffraction is the bending of light waves upon encountering obstacles or slits and resulting in typical patterns. Interference is the combination of several waves of light, has constructive or destructive effects that find extensive uses in holography and interferometry. Scattering is a change in the path of light brought about by the interaction of the light with particles or surface irregularities, used extensively in communications through optical systems and the effects observed in the atmosphere.

In industrial optical systems, numerous optical effects are responsible for sophisticated functionalities. Polarization [32] is one such effect which is critical in display technology and in optical filters. Coherence [33] determines the phase relationships in light waves, which is an important basis for industrial laser systems. Evanescent

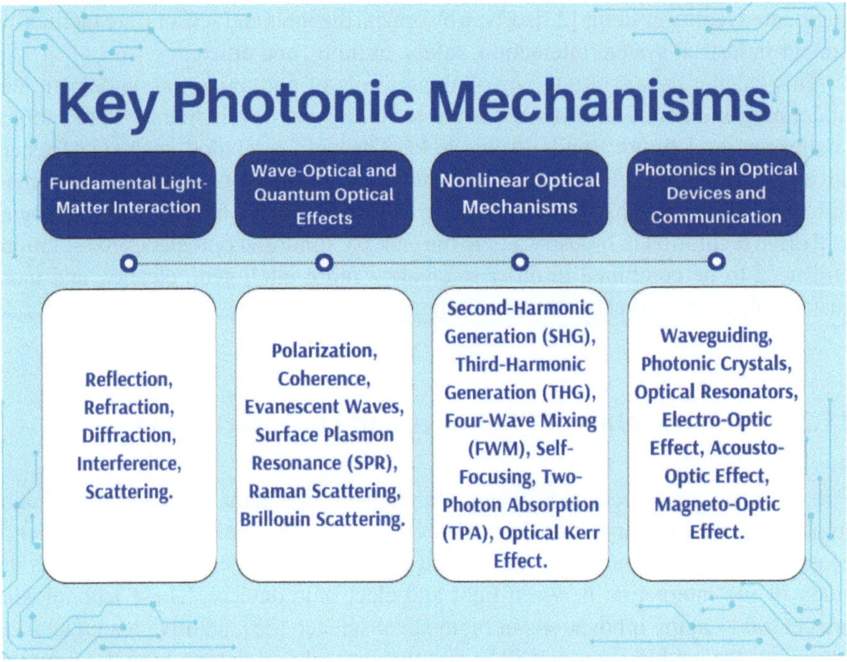


**Fig. 11.1** Key mechanisms in photonics, including fundamental light-matter interactions, wave-optical and quantum-optical effects, nonlinear optical phenomena, and device-level optical processes

waves [34, 35] which exponentially decay near interfaces are used for optical waveguides [36] and biosensors [37]. Surface Plasmon Resonance (SPR) [38] enables the coupling of light with electron oscillations at a metal–dielectric interface, commonly employed in chemical and biological sensing [39]. Raman and Brillouin scattering [40] are based on inelastic interactions between light and molecular vibrations or acoustic phonons, respectively, providing insight into microwave photonics and optical signal processing [41]. In addition, nonlinear optical processes like second-harmonic generation (SHG) [42], third-harmonic generation (THG) [43], and four-wave mixing (FWM) [44] play important roles in frequency conversion [45] and ultra-fast photonics, whereas phenomena like self-focusing [46], two-photon absorption (TPA), and the optical Kerr effect contributes to improved optical modulation and switching performances [47, 48].

Optoelectronic processes that form the basis of Industry 5.0 are depicted in Fig. 11.2. Among the significant processes in optoelectronics, photoelectric effect [49] that controls the release of electrons upon exposure to light finds extensive application in high-sensitivity photodetectors [50], facilitating sensing of the environment and atmosphere and processing of information [51]. Likewise, electroluminescence [52], the technology powering LEDs and OLEDs [53], is revolutionizing industrial light solutions through energy-saving, high-brightness displays, and smart

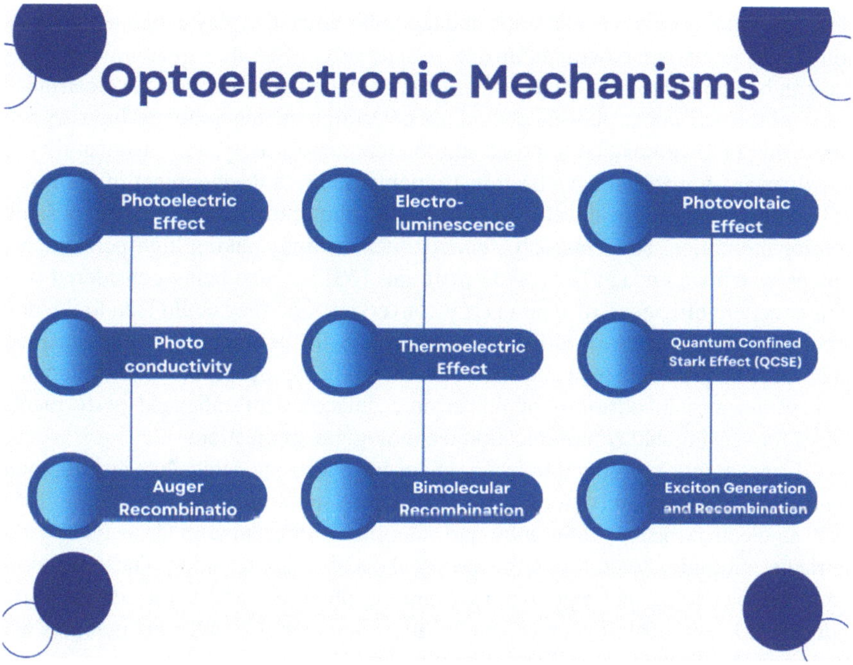


**Fig. 11.2** Overview of key optoelectronic mechanisms fundamental to the operation of devices for Industry 5.0

visual interfaces in intelligent production settings. The technologies are sustainable and energy-efficient [54] and respond to Industry 5.0's emphasis on human-centred, eco-friendly innovation.

The photovoltaic effect [55] is essential in green energy solutions for smart manufacturing, since solar cells powers production lines for quality monitoring and control [56]. In addition, photoconductivity [57] is used in high-performance sensing, where materials become more electrically conductive when light is applied. This improves accuracy in optical sensors used for defect analysis and quality monitoring. Thermoelectric effect, which allows for the conversion of waste heat into electrical energy through light absorption, is crucial in self-sustaining industrial systems that reduce energy waste [58]. Moreover, Quantum Confined Stark Effect (QCSE) plays a crucial role in electro-absorption modulators, maximizing high-speed optical communication networks that are critical to applications in datacomms and linear optical quantum computing [59].

Recombination processes are important for development of next-generation optoelectronic devices. Auger recombination [60] is a non-radiative phenomenon in which the recombination energy is transferred to another carrier, affects the efficiency of semiconductor LEDs and lasers used in industrial manufacturing. In addition, bimolecular recombination [61], a process where electrons and holes recombine and emit photons, is at the heart of high-efficiency displays and light sources in smart factories. Finally, exciton generation and recombination also play a role in the formation of new semiconductor quantum dots [62] for next-generation quantum-dot LEDs that can be applied to next-generation display technologies and flexible electronics. These optoelectronic processes as a whole determine the landscape of Industry 5.0, supporting innovations that improve automation, connectivity, and sustainability.

Advances in material have been instrumental in swift development of photonics and optoelectronics. The development of silicon photonics [13] has made it possible to integrate optical elements onto semiconductor chips, making high-performance optical systems compact. Perovskite materials [63] are also being considered with their excellent efficiency in solar energy conversion, and they could lead to the next generation of solar cells. In addition, discovery and development of two-dimensional materials like graphene [64] and transition metal dichalcogenides [65] have opened up new avenues in ultra-fast optoelectronic devices with enhanced performance useful for sensing and communication technological applications.

Advancements in production technologies have also accelerated the development of new age photonics and optoelectronic devices. Nanofabrication technologies [66], such as electron beam lithography and nanoimprinting, have made it possible to fabricate photonic structures at the nanoscale with high accuracy. These developments are providing the impetus for integrating photonics and optoelectronics into Industry 5.0, where intelligent, energy-saving, and human-friendly technologies will determine the direction of industrial applications.

## 11.3 Key Technological Trends in Photonics and Optoelectronics for Industry 5.0

The transformation to Industry 5.0 is all about human-oriented, sustainable, and intelligent industrial systems, being a vital paradigm shift from the automation-oriented approach of Industry 4.0. Photonics and optoelectronics form the core of this transformation, enabling the growth of new capabilities in sensing, imaging, communications, and energy efficiency as detailed in Sect. 11.2. These are now increasingly supported by advances in artificial intelligence (AI), quantum mechanics, and integrated photonics, to create an era in which industries collaborate with increased precision, efficiency, and with a lower environmental footprint (Fig. 11.3).

### *11.3.1 Smart Photonic Sensors and Imaging*

One of the significant advances in Industry 5.0 has been the development of smart photonic sensors [67] and imaging technology. High-sensitivity optical sensors are a

**Fig. 11.3** Key technological trends in photonics and optoelectronics driving advancements in Industry 5.0

major part of industrial real-time monitoring and quality control. Light-based detection methods are employed by photonic sensors, as opposed to other sensors, in order to achieve higher sensitivity, accuracy, and speed.

Artificial intelligence-powered image processing along with photonic sensing allows automation of complex visual data. Hyperspectral imaging, which captures a wide range of wavelengths beyond the visible spectrum, is being increasingly used in conjunction with AI algorithms to enhance defect detection and process monitoring in semiconductor fabrication, food processing, and medical imaging [68, 69]. With the use of AI for real-time decision-making, these imaging platforms enable predictive maintenance and improves overall operation efficiency.

In addition, the integration of photonic sensors and artificial intelligence [70] enables industries to identify slight changes in materials and processes. This is important in industries such as aerospace and automotive production, where slight imperfections can cause major performance problems.

### *11.3.2 Integrated Photonics and Silicon Photonics*

Another important trend fuelling the growth of Industry 5.0 is the miniaturization of photonic devices. Silicon photonics [13] and integrated photonics [71] are making it possible to create high-performance optical systems that are compact in size and can be easily interfaced with current industrial and communication networks. These technologies find important application in high-speed data transmission, where optical interconnects are being used in place of conventional electronic circuits to cater to the increasing demands for higher speed and lower energy consumption in communication systems.

Photonic integrated circuits (PICs) [72, 73] are leaders in this revolution, integrating a number of photonic devices like lasers, modulators, and detectors on a single chip. PICs have advantages in size miniaturization, energy efficiency, and scalability over traditional electronic solutions which makes them suitable for applications in industry.

Optical sensors are being integrated into manufacturing machines to give instant feedback on the performance of the machine, allowing for adaptive process control and energy savings. Silicon photonic manufacturing [74] is being used in smart factories to enable ultra-high-speed data exchange among various components to make industrial systems more responsive.

### *11.3.3 Quantum Photonics*

Quantum photonics [75] is a new area of research which finds applications in secure communications, computing, and imaging. Through the use of quantum principles, quantum photonic technologies are achieving new possibilities unachievable by traditional classical systems. In quantum communication, one of the most profound applications is quantum key distribution (QKD) which allows for ultra-secure data transfer using the inherent characteristics of photons. This technology will revolutionize cybersecurity in Industry 5.0 by offering unbreakable encryption for industrial networks.

Quantum sensors which are based on entangled photons, have potential to be used in sensitive measurement devices, allowing for accurate detection of structural flaws. The application of such sensors is being researched for non-destructive testing and high-resolution medical imaging [76].

### *11.3.4 AI and Machine Learning in Photonics*

The integration of photonics and AI is propelling innovation for industrial use [77] by using AI-aided design and optimization of photonic devices. AI algorithms are being employed to model and optimize photonic devices, minimizing development time and enhancing performance. AI-based deep learning and machine learning simulations are being used to predict the performance of intricate optical systems which enables engineers to tailor parameters prior to physical prototyping.

Predictive maintenance is another sector where optical sensing and AI are having a substantial impact [78]. Optical sensors can be integrated into industrial equipment to continuously measure vibrations, temperature, and strain and produce huge amounts of data that AI algorithms process to forecast possible failures. Industries can then carry out maintenance prior to a breakdown by detecting early indications of wear and tear, reducing downtime and operating expenses.

Deep learning methods are also being combined with photonic imaging systems to improve defect detection and improve process control in manufacturing. In semiconductor production manufacturing, visual optical inspection systems [79] are being used to detect microscopic defects in wafers more accurately than with conventional techniques.

### *11.3.5 Energy-Efficient and Sustainable Optoelectronics*

Industry 5.0 technologies are aimed to focus more on sustainability, energy-efficient photonic, and optoelectronic technologies. The most promising aspect of this technology is the application of low-power photonic components to industrial processes. Optical computing [10], for instance, provides a competent substitute for traditional electronic processors by saving massive amounts of energy without compromising processing speeds. The use of solar photovoltaic optoelectronic technology is also key to sustainable manufacturing processes. Modern solar cells based on perovskite materials and multi-junction designs provide enhanced efficiencies in energy conversion, allowing factories to increasingly utilize renewable sources.

Green photonics [17], concentrating on the application of light-based technology for environmental sustainability, is becoming a central area of research. Precision manufacturing based on lasers, for example, reduces material loss and energy usage relative to traditional machining. Optical recycling systems are also under investigation for enhanced waste reduction in industries by means of photonic sorting methods to sort recyclable materials more effectively.

Another key use of sustainable optoelectronics [80] is in smart lighting systems, where low-energy LEDs are coupled with AI-based lighting control software. Such systems control illumination in relation to occupancy as well as to the ambient light levels which minimizes wastage of energy in commercial as well as industrial environments. As the focus on carbon footprint reduction intensifies, photonic and optoelectronic technologies will be at the forefront of developing greener and more sustainable industrial ecosystems. The integration of photonics and optoelectronics into Industry 5.0 is revolutionizing the way industries work, concentrating on efficiency as well as sustainability.

## 11.4 Applications in Industry 5.0

As the industrial ecosystem transitions towards more intelligent, human-centric, and sustainable solutions, optoelectronic and photonic technologies are playing a crucial role in enabling advancements in manufacturing, communication, health care, and

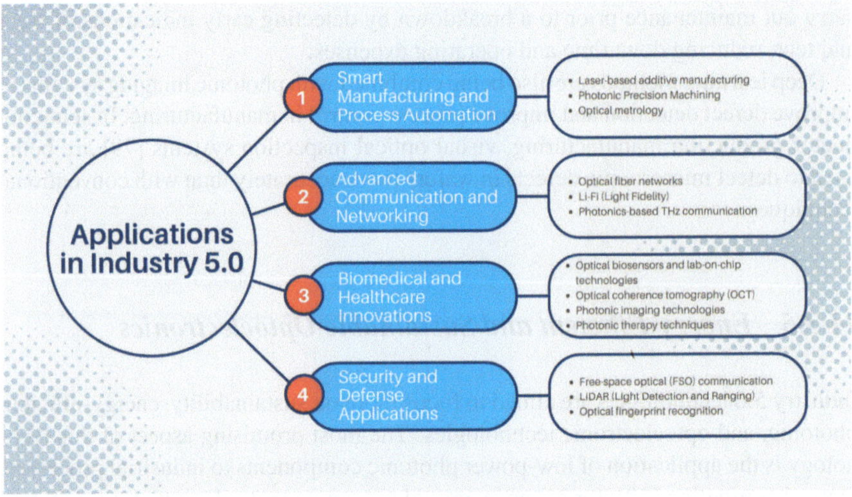


**Fig. 11.4** Applications of photonics and optoelectronics in Industry 5.0

security. With their ability to enhance precision, efficiency, and data processing capabilities, photonics-based innovations are transforming industries, making them more adaptive and interconnected. This section explores some of the key applications of photonics and optoelectronics in Industry 5.0 (Fig. 11.4).

### *11.4.1 Smart Manufacturing and Process Automation*

The convergence of photonic technologies with smart manufacturing has opened doors to extremely accurate, efficient, and automated processes in industry. Laser-based additive manufacturing [81], also known as 3D printing, has acquired a lot of momentum because of its capability for the production of intricate structures at high accuracy levels. In contrast to conventional approaches to manufacturing, laser-based manufacturing techniques allow instant prototyping, material optimization, and minimizes wastage. Aerospace, automotive, and biomedical engineering sectors are using photonic-powered additive manufacturing to fabricate light-weight yet robust parts with tailor-made designs.

Outside of additive manufacturing, photonic precision machining is transforming material processing with high-speed, contactless fabrication methods. Laser cutting, drilling, and welding offer unmatched accuracy, making them a necessity for applications that demand detailed detailing and low thermal damage. These methods are commonly applied in semiconductor production, electronics manufacturing, and high-precision engineering applications.

Optical metrology [82] is yet another critical use of photonics in smart manufacturing. Real-time monitoring of the industrial process and quality control depends on optical measurement technologies like interferometry, laser scanning, and spectroscopic analysis. Optical measurement systems make non-contact high-resolution inspections of components under production possible, thereby ensuring defect-free production and strict compliance with quality standards. Optical metrology is especially beneficial in semiconductor manufacturing, where microscopic precision is necessary to manufacture advanced microchips and circuits.

### *11.4.2 Advanced Communication and Networking*

The growing need for super-speed data transfer in Industry 5.0 has promoted the universal use of optical communication technologies. Optical fibre networks [83] offer low-latency, high-speed, and interference-free data transfer and thus are the mainstay of industrial automation and digitalization. As compared to conventional copper-based communication networks, fibre-optic networks provide better bandwidth, ensuring unobstructed connectivity between industrial units and supply chains.

Light Fidelity (Li-Fi) [84] is the next potential challenger to Wi-Fi, using visible light for wireless data transfer. By modulating LED light signals, Li-Fi provides high-speed, secure, and interference-immune communication that is best suited for industrial environments where electromagnetic interference is a key issue. The technology finds widespread application in smart factories, healthcare centres, and defence systems, where data security and real-time connectivity are of utmost importance.

Another important development brought about by photonics is the advancement of communication networks beyond 5G. The advent of 6G networks [85] and even higher is projected to be heavily dependent on terahertz (THz) and visible light communications technologies. THz waves have very high data rates and are ideal for ultra-high-speed wireless networks and future industrial automation. Photonics-driven THz communication can support real-time connectivity in smart manufacturing, autonomous systems, and industrial IoT applications. Furthermore, visible light communication (VLC) is also investigated for wireless data exchange in connected environments, further improving the speed and reliability of industrial networks.

### *11.4.3 Biomedical and Healthcare Innovations*

Photonics is revolutionizing the biomedical and healthcare industries, enabling next-generation diagnostic and therapeutic solutions. Optical biosensors [86] and lab-on-chip technology [87] are transforming personalized medicine through rapid and precise detection of disease. These devices leverage photonic concepts like fluorescence, Raman spectroscopy, and plasmonic sensing to probe biomolecular interactions in real time. Lab-on-chip platforms combine microfluidics and optics, enabling point-of-care diagnostics with low sample volumes. They find their greatest use in the identification of infectious diseases, cancer markers, and metabolic diseases.

Non-invasive photonics-based diagnostic methods are dramatically enhancing the care of patients. High-resolution imaging technology optical coherence tomography (OCT) [88] is commonly applied for retinal imaging, dermatology, and cardiovascular diagnosis. In a similar way, near-infrared spectroscopy (NIRS) [89] is utilized for tissue oxygen monitoring, contributing to brain function analysis and neonatal management. Photonic imaging technologies like multiphoton microscopy and hyperspectral imaging are adding precision to medical diagnoses, allowing the early detection of diseases with increased accuracy.

Telemedicine has also benefited from photonics advancements, particularly in remote health monitoring and digital pathology. Wearable optical sensors [90] can continuously monitor physiological parameters such as heart rate, blood oxygen, and glucose level, allowing real-time monitoring of health.

Photonic therapy methods, which include laser-based treatments and photodynamic therapy (PDT), are extensively being applied to non-invasive medical treatments. Lasers are used in ophthalmology, dermatology, and dental operations for targeted tissue ablation and regeneration. PDT, which consists of the activation of photosensitizing agents with a particular light wavelength, is a highly efficient form of treatment for some cancers and bacterial infections. These photonics-based treatments provide low side effects and enhanced treatment outcomes, thereby driving personalized healthcare.

### *11.4.4 Security and Defence Applications*

Photonics and optoelectronics are now crucial components of security and defence technology, providing sophisticated solutions for surveillance, communications, and threat detection. Quantum cryptography based on the foundations of quantum mechanics, is transforming secure communication by supplying encryption techniques that are theoretically inviolable. Quantum key distribution (QKD) allows ultra-secure data transmission, and it is very useful in military communications, financial transactions, and protecting critical infrastructure.

Secure optical communications is another important defence application of photonics. Free-space optical (FSO) communication systems [91] employ laser beams to communicate over long distances wirelessly. These systems provide high-speed, secure, and interference-immune communication and, therefore, are well-suited for military and aerospace missions. FSO technology finds extensive use in situations where conventional communication networks are not practical, for example, in underwater and satellite-based communication systems.

Light Detection and Ranging (LiDAR) [92] technology is a critical tool used in defence and autonomous systems. LiDAR uses laser pulses to accurately measure distances, thus allowing real-time mapping, target recognition, and obstacle detection. In defence systems, LiDAR is utilized for battlefield intelligence gathering, surveillance, and missile guidance systems. In the robotics and automotive sector, LiDAR is utilized to facilitate autonomous navigation by creating accurate spatial awareness and object detection.

Photonics-based radar systems are also increasing security and surveillance performance. Unlike radio wave-based conventional radar systems, photonic radar uses optical signals to detect and track objects with greater resolution and accuracy. These systems are very effective in detecting stealth aircraft, coastal security surveillance, and intrusions into sensitive areas without authorization. AI-photonics integration in radar technology further improves target classification and threat evaluation.

Besides defence, photonic security technologies [93] are also being used in biometric verification and counterfeiting detection. Optical fingerprint recognition, iris scanning, and facial recognition systems use photonic imaging methods to ensure secure and authentic identification. Holographic and fluorescence-based anti-counterfeiting techniques are also being used to secure valuable documents, currency, and consumer products from counterfeiting.

The applications of photonics and optoelectronics in Industry 5.0 go far beyond the usual, spearheading innovation in many fields such as manufacturing, communications, medicine, and security. Intelligent photonic technologies are empowering precision manufacturing, real-time quality monitoring, and sophisticated automation. Optical communication technology like fibre optic, Li-Fi, and THz communication is transforming data communication, enabling ultra-high-speed industrial networks. In medicine, photonic biosensors, imaging, and therapeutic applications are improving diagnostics and treatment outcomes. At the same time, security and defence photonics-based applications, such as quantum cryptography, LiDAR, and photonic radar [94], are making systems more secure and resilient. As Industry 5.0 develops, the combination of these sophisticated photonic technologies will accelerate progress further, creating a future that is more connected, intelligent, and sustainable.

## 11.5 Challenges and Future Directions

As photonics and optoelectronics evolve in Industry 5.0, a number of challenges need to be overcome to make them ubiquitous. Scalability and cost are major issues, as the production of high-precision optical devices is expensive and labour-intensive. Cost-effective manufacturing methods like wafer-scale fabrication, 3D printing, and integration of photonics with semiconductor processes are areas of research. Moreover, hybrid photonic-electronic systems come with alignment, signal processing, and thermal management complexities. To address these issues, emerging packaging solutions, silicon photonics, and artificial intelligence-assisted design optimization are being investigated to increase compatibility and efficiency.

Apart from technical issues, ethical and regulatory issues need to be resolved, especially in surveillance, health care, and secure communications. Optical sensors and imaging solutions pose privacy questions, requiring stringent regulatory environments. In health care, the safety and reliability of photonic medical equipment are essential, as well as adherence to global standards. The future will see research in quantum photonics, energy-efficient optics, and AI-optimized performance driving future advances. The combination of photonics with neuromorphic computing and bio-inspired optical systems will further increase its application in smart industries, making it more efficient, intelligent, and sustainable in the emerging digital world.

## 11.6 Conclusion

Photonics and optoelectronics are central enablers of Industry 5.0, powering intelligent, sustainable, and human-centred innovation in manufacturing, health care, communication, and security applications. As the focus is shifting from conventional electronics to high-speed, low-energy photonic technologies, applications such as optical sensors, quantum imaging, Li-Fi, and PICs are transforming industrial processes. AI and machine learning further optimize photonic system design to support predictive analytics and automation. Applications vary from laser-based fabrication and non-invasive diagnostics to secure quantum communication and high-speed optical data transmission, with critical applications in next-generation communication systems, such as 6G networks.

In spite of these developments, scalability, cost minimization, and integration with current electronic systems are still challenges. The need for sustainable photonic devices and ethical standards in security, medicine, and communication is increasing. Quantum photonics breakthroughs, AI-optimized photonic design, and the intersection with neuromorphic computing and bio-inspired optics will define the next era of Industry 5.0. As studies continue, photonics will also propel automation, connectivity, and intelligence, making Industry 5.0 not only a technological revolution but also a transformational and sustainable industrial age.

## References


1. Leng J, Sha W, Wang B, Zheng P, Zhuang C, Liu Q, Wuest T, Mourtzis D, Wang L. Industry 5.0: prospect and retrospect. J Manuf Syst. 2022;65:279–95.
2. Xu X, Lu Y, Vogel-Heuser B, Wang L. Industry 4.0 and Industry 5.0—inception, conception and perception. J Manuf Syst. 2021;61:530–5.
3. Massaro A. Advanced electronic and optoelectronic sensors, applications, modelling and industry 5.0 perspectives. Appl Sci. 2023;13(7):4582.
4. Chen Z, Segev M. Highlighting photonics: looking into the next decade. ELight. 2021;1(1):2.
5. Parker MA. Physics of optoelectronics. CRC Press; 2018.
6. Yang J, Liu Y, Morgan PL. Human-machine interaction towards Industry 5.0: human-centric smart manufacturing. Digit Eng. 2024:100013.
7. Aheleroff S, Huang H, Xu X, Zhong RY. Toward sustainability and resilience with Industry 4.0 and Industry 5.0. Front Manuf Technol. 2022;2:951643.

8. Butt MA, Mateos X, Piramidowicz R. Photonics sensors: a perspective on current advancements, emerging challenges, and potential solutions. Phys Lett A. 2024;516: 129633.
9. Kazanskiy NL, Butt MA, Khonina SN. Optical computing: status and perspectives. Nanomaterials. 2022;12(13):2171.
10. McMahon PL. The physics of optical computing. Nat Rev Phys. 2023;5(12):717–34.
11. Politi A, Matthews JC, Thompson MG, O'Brien JL. Integrated quantum photonics. IEEE J Sel Top Quantum Electron. 2009;15(6):1673–84.
12. Chen Z, Obaid SN, Lu L. Recent advances in organic optoelectronic devices for biomedical applications. Opt Mater Exp. 2019;9(9):3843–56.
13. Jalali B, Fathpour S. Silicon photonics. J Lightwave Technol. 2006;24(12):4600–15.
14. Reed GT. Silicon photonics: the state of the art.
15. Uddin S, Shareef H, Mohamed A, Hannan MA, Mohamed K. LEDs as energy efficient lighting systems: a detail review. In: 2011 IEEE student conference on research and development 2011 (pp. 468–472). IEEE.
16. Quan F. Green photonics. J Opt. 2012;14(2): 024001.
17. Eisenstein G, Bimberg D, editors. Green photonics and electronics. Springer International Publishing; 2017.
18. Parida B, Iniyan S, Goic R. A review of solar photovoltaic technologies. Renew Sustain Energy Rev. 2011;15(3):1625–36.
19. Solanki CS. Solar photovoltaics: fundamentals, technologies and applications. Phi Learning Pvt. Ltd.; 2015.
20. Victoria M, Haegel N, Peters IM, Sinton R, Jäger-Waldau A, Del Cañizo C, Breyer C, Stocks M, Blakers A, Kaizuka I, Komoto K. Solar photovoltaics is ready to power a sustainable future. Joule. 2021;5(5):1041–56.
21. Nahavandi S. Industry 5.0—a human-centric solution. Sustainability. 2019;11(16):4371.
22. Lu Y, Zheng H, Chand S, Xia W, Liu Z, Xu X, Wang L, Qin Z, Bao J. Outlook on human-centric manufacturing towards Industry 5.0. J Manuf Syst. 2022;62:612–27.
23. Goel R, Baral SK, Mishra T, Jain V, editors. Augmented and virtual reality in Industry 5.0. Walter de Gruyter GmbH & Co KG; 2023.
24. Kasap SO. Optoelectronics and photonics. Pearson Education UK; 2013.
25. Vo-Dinh T, editor. Biomedical photonics handbook: biomedical diagnostics. CRC Press; 2014.
26. Camposeo A, Persano L, Farsari M, Pisignano D. Additive manufacturing: applications and directions in photonics and optoelectronics. Adv Opt Mater. 2019;7(1):1800419.
27. Suki T, Mondal A, Mahajan C, Joglekar PV. Sustainable photonics for energy harvesting and beyond, from solar cells to OLEDs, efficient communication, and smart windows. Interactions. 2024;245(1):130.
28. Born M, Wolf E. Principles of optics: electromagnetic theory of propagation, interference and diffraction of light. Elsevier; 2013.
29. Zhou X, Yi D, Chan DW, Tsang HK. Silicon photonics for high-speed communications and photonic signal processing. NPJ Nanophotonics. 2024;1(1):27.
30. Dornfeld D, Lee DE. Sensors for precision manufacturing. US: Springer; 2008.
31. Meschede D. Optics, light and lasers: the practical approach to modern aspects of photonics and laser physics. John Wiley & Sons; 2017.
32. Goldstein D, Goldstein DH. Polarized Light, revised and expanded. CRC Press; 2003.
33. Steinfeld J, editor. Laser and coherence spectroscopy. Springer Science & Business Media; 2013.
34. Felsen LB. Evanescent waves. J Opt Soc Am. 1976;66(8):751–60.
35. De Fornel F. Evanescent waves: from Newtonian optics to atomic optics. Springer Science & Business Media; 2001.
36. Povinelli ML, Lončar M, Ibanescu M, Smythe EJ, Johnson SG, Capasso F, Joannopoulos JD. Evanescent-wave bonding between optical waveguides. Opt Lett. 2005;30(22):3042–4.
37. Taitt CR, Anderson GP, Ligler FS. Evanescent wave fluorescence biosensors. Biosens Bioelectron. 2005;20(12):2470–87.

38. Homola J, Piliarik M. Surface plasmon resonance (SPR) sensors. Berlin Heidelberg: Springer; 2006.
39. Homola J. Surface plasmon resonance sensors for detection of chemical and biological species. Chem Rev. 2008;108(2):462–93.
40. Reintjes JF. Stimulated Raman and Brillouin scattering. In: CRC handbook of laser science and technology supplement 2 2020 (pp. 334–364). CRC Press.
41. Preussler S, Schneider T. Stimulated Brillouin scattering gain bandwidth reduction and applications in microwave photonics and optical signal processing. Opt Eng. 2016;55(3):031110.
42. Chen J, Hu CL, Kong F, Mao JG. High-performance second-harmonic-generation (SHG) materials: new developments and new strategies. Acc Chem Res. 2021;54(12):2775–83.
43. Markowicz PP, Tiryaki H, Pudavar H, Prasad PN, Lepeshkin NN, Boyd RW. Dramatic enhancement of third-harmonic generation in three-dimensional photonic crystals. Phys Rev Lett. 2004;92(8): 083903.
44. Konorov SO, Fedotov AB, Zheltikov AM. Enhanced four-wave mixing in a hollow-core photonic-crystal fiber. Opt Lett. 2003;28(16):1448–50.
45. Wang C, Wen Y, Sun J, Zhou J. Recent progress on optical frequency conversion in nonlinear metasurfaces and nanophotonics. ES Mater Manuf. 2022;17(2):1–3.
46. Shen YR. Self-focusing: experimental. Prog Quantum Electron. 1975;4:1–34.
47. Yin L, Agrawal GP. Impact of two-photon absorption on self-phase modulation in silicon waveguides. Opt Lett. 2007;32(14):2031–3.
48. England D, Bouchard F, Fenwick K, Bonsma-Fisher K, Zhang Y, Bustard PJ, Sussman BJ. Perspectives on all-optical Kerr switching for quantum optical applications. Appl Phys Lett. 2021;119(16).
49. Einstein A. On a heuristic point of view concerning the production and transformation of light. Ann Phys. 1905;17(132):1–6.
50. Li G, Wang Y, Huang L, Sun W. Research progress of high-sensitivity perovskite photodetectors: a review of photodetectors: noise, structure, and materials. ACS Appl Electron Mater. 2022;4(4):1485–505.
51. Ma T, Xue N, Muhammad A, Fang G, Yan J, Chen R, Sun J, Sun X. Recent progress in photodetectors: from materials to structures and applications. Micromachines. 2024;15(10):1249.
52. Kafafi ZH. Organic electroluminescence. CRC Press; 2018.
53. Huang Y, Hsiang EL, Deng MY, Wu ST. Mini-LED, Micro-LED and OLED displays: present status and future perspectives. Light: Sci Appl. 2020;9(1):105.
54. Thejokalyani N, Dhoble SJ. Novel approaches for energy efficient solid state lighting by RGB organic light emitting diodes–a review. Renew Sustain Energy Rev. 2014;32:448–67.
55. Dai Z, Rappe AM. Recent progress in the theory of bulk photovoltaic effect. Chem Phys Rev. 2023;4(1).
56. Pei FQ, Tong YF, Yuan MH, Ding K, Chen XH. The digital twin of the quality monitoring and control in the series solar cell production line. J Manuf Syst. 2021;59:127–37.
57. Sumanth A, Ganapathi KL, Rao MR, Dixit T. A review on realizing the modern optoelectronic applications through persistent photoconductivity. J Phys D Appl Phys. 2022;55(39): 393001.
58. Singh R, Dogra S, Dixit S, Vatin NI, Bhardwaj R, Sundramoorthy AK, Perera HC, Patole SP, Mishra RK, Arya S. Advancements in thermoelectric materials for efficient waste heat recovery and renewable energy generation. Hybrid Adv. 2024:100176.
59. Murphy T, Broderick CA, Peters FH, O'Reilly EP. Lateral quantum-confined stark effect for integrated quantum dot electroabsorption modulators. IEEE J Sel Top Quantum Electronics. 2025.
60. Al-Mousoi AK, Mohammed MK, Kumar A, Pandey R, Madan J, Dastan D, Hossain MK, Sakthivel P, Yaseen ZM. Understanding Auger recombination in perovskite solar cells. Phys Chem Chem Phys. 2023;25(24):16459–68.
61. Hossain K, Sivadas D, Kabra D, Nair PR. Perovskite solar cells dominated by bimolecular recombination—how far is the radiative limit? ACS Energy Lett. 2024;9(5):2310–7.
62. Kanemitsu Y. Multiple exciton generation and recombination in carbon nanotubes and nanocrystals. Acc Chem Res. 2013;46(6):1358–66.

63. Jung HS, Park NG. Perovskite solar cells: from materials to devices. Small. 2015;11(1):10–25.
64. Bonaccorso F, Sun Z, Hasan T, Ferrari AC. Graphene photonics and optoelectronics. Nat Photonics. 2010;4(9):611–22.
65. Wang QH, Kalantar-Zadeh K, Kis A, Coleman JN, Strano MS. Electronics and optoelectronics of two-dimensional transition metal dichalcogenides. Nat Nanotechnol. 2012;7(11):699–712.
66. Cui Z. Nanofabrication. US: Springer; 2008.
67. Passaro VM, De Tullio C, Troia B, La Notte M, Giannoccaro G, De Leonardis F. Recent advances in integrated photonic sensors. Sensors. 2012;12(11):15558–98.
68. Khanam R, Hussain M, Hill R, Allen P. A comprehensive review of convolutional neural networks for defect detection in industrial applications. IEEE Access. 2024.
69. Zheng X, Zheng S, Kong Y, Chen J. Recent advances in surface defect inspection of industrial products using deep learning techniques. Int J Adv Manuf Technol. 2021;113:35–58.
70. Zhang S, Wei S, Liu Z, Li T, Li C, Huang XL, Wang C, Xie Z, Al-Hartomy OA, Al-Ghamdi AA, Wageh S. The rise of AI optoelectronic sensors: from nanomaterial synthesis, device design to practical application. Mater Today Phys. 2022;27: 100812.
71. Pollock C, Lipson M. Integrated photonics. 2003.
72. Nagarajan R, Joyner CH, Schneider RP, Bostak JS, Butrie T, Dentai AG, Dominic VG, Evans PW, Kato M, Kauffman M, Lambert DJ. Large-scale photonic integrated circuits. IEEE J Sel Top Quantum Electron. 2005;11(1):50–65.
73. Kish F, Lal V, Evans P, Corzine SW, Ziari M, Butrie T, Reffle M, Tsai HS, Dentai A, Pleumeekers J, Missey M. System-on-chip photonic integrated circuits. IEEE J Sel Top Quantum Electron. 2017;24(1):1–20.
74. Zortman WA, Trotter DC, Watts MR. Silicon photonics manufacturing. Opt Express. 2010;18(23):23598–607.
75. Silverstone JW, Bonneau D, O'Brien JL, Thompson MG. Silicon quantum photonics. IEEE J Sel Top Quantum Electron. 2016;22(6):390–402.
76. Aslam N, Zhou H, Urbach EK, Turner MJ, Walsworth RL, Lukin MD, Park H. Quantum sensors for biomedical applications. Nat Rev Phys. 2023;5(3):157–69.
77. Mahmoud MG, Hares AS, Hameed MF, El-Azab MS, Obayya SS. AI-driven photonics: unleashing the power of AI to disrupt the future of photonics. APL Photonics. 2024;9(8).
78. Murtaza AA, Saher A, Zafar MH, Moosavi SK, Aftab MF, Sanfilippo F. Paradigm shift for predictive maintenance and condition monitoring from Industry 4.0 to Industry 5.0: a systematic review, challenges and case study. Results Eng. 2024:102935.
79. Huang SH, Pan YC. Automated visual inspection in the semiconductor industry: a survey. Comput Ind. 2015;1(66):1.
80. Bhandari S, Mondal D, Nataraj SK, Balakrishna RG. Biomolecule-derived quantum dots for sustainable optoelectronics. Nanoscale Adv. 2019;1(3):913–36.
81. Hu D, Kovacevic R. Sensing, modeling and control for laser-based additive manufacturing. Int J Mach Tools Manuf. 2003;43(1):51–60.
82. Gåsvik KJ. Optical metrology. John Wiley & Sons; 2003.
83. Winzer PJ. Scaling optical fiber networks: challenges and solutions. Opt Photonics News. 2015;26(3):28–35.
84. Khanda D, Jain S. Li-fi (light fidelity): the future technology in wireless communication. Int J Inf Comput Technol. ISSN. 2014:0974-2239.
85. Giordani M, Polese M, Mezzavilla M, Rangan S, Zorzi M. Toward 6G networks: use cases and technologies. IEEE Commun Mag. 2020;58(3):55–61.
86. Damborský P, Švitel J, Katrlík J. Optical biosensors. Essays Biochem. 2016;60(1):91–100.
87. García-Meca C, Lechago S, Brimont A, Griol A, Mas S, Sánchez L, Bellieres L, Losilla NS, Martí J. On-chip wireless silicon photonics: from reconfigurable interconnects to lab-on-chip devices. Light: Sci Appl. 2017;6(9):e17053.
88. Schmitt JM. Optical coherence tomography (OCT): a review. IEEE J Sel Top Quantum Electron. 2002;5(4):1205–15.
89. Ferrari M, Mottola L, Quaresima V. Principles, techniques, and limitations of near infrared spectroscopy. Can J Appl Physiol. 2004;29(4):463–87.

90. Ballard ZS, Ozcan A. Wearable optical sensors. Mobile health: Sensors, analytic methods, and applications. 2017:313–42.
91. Kaushal H, Jain VK, Kar S. Free space optical communication. New Delhi: Springer India; 2017.
92. Reutebuch SE, Andersen HE, McGaughey RJ. Light detection and ranging (LIDAR): an emerging tool for multiple resource inventory. J Forest. 2005;103(6):286–92.
93. Cutolo A. Photonics for safety and security. World Scientific; 2013.
94. De S, Bazil Raj AA. A survey on photonics technologies for radar applications. J Opt. 2023;52(1):90–119.

MIX
Papier aus verantwortungsvollen Quellen
Paper from responsible sources
FSC® C105338

If you have any concerns about our products,
you can contact us on
**ProductSafety@springernature.com**

In case Publisher is established outside the EU,
the EU authorized representative is:
**Springer Nature Customer Service Center GmbH**
**Europaplatz 3, 69115 Heidelberg, Germany**

Printed by Libri Plureos GmbH
in Hamburg, Germany